THE CRANBOURNE METEORITE

THE
CRANBOURNE METEORITE

SEAN MURPHY

© Sean Murphy 2023

First published 2023 by
Australian Scholarly Publishing Pty Ltd
7 Lt Lothian St North
North Melbourne, Victoria 3051

tel: 61 3 93296963
enquiry@scholarly.info / www.scholarly.com

ISBN: 978-1-923068-72-8

ALL RIGHTS RESERVED

Cover image: Museums Victoria / Photographer: Rodney Start
Cover design: David Morgan

For my parents, Maureen and Jim
In memoriam

Contents

Introduction ix

1: A Memo of all the Facts 1

2: Two Masses of Malleable Iron 11

3: Our Colonial Crystal Palace 23

4: A Somewhat Eccentric Man 36

5: This Mysterious Visitant 48

6: Divisions over Division 63

7: Of Mutual Interest and Support 75

8: This Out-of-the-way Part of the World 86

9: So Magnificent a Cabinet-Piece 96

10: Without a Violation of Good Faith 104

11: The Honour and the Gain 118

12: Epilogue 137

Notes 149

Index 157

Main Characters

Ferdinand Mueller	Victorian government botanist, Director Royal Botanic Gardens, Melbourne
Frederick McCoy	University of Melbourne professor, Director National Museum of Victoria
James Bruce	Settler
Hugh McKay	Settler
Augustus Abel	Mineralogist
James Lineham	Settler
Nevil Story Maskelyne	Keeper of Minerals, British Museum
Sir Henry Barkly	Governor of Victoria
Edmund FitzGibbon	Melbourne Town Clerk
Alfred Selwyn	Director Geological Survey of Victoria
Robert Brough Smyth	Secretary of Mines, Victoria
Richard Evans	Postmaster-general, Victoria
Richard Daintree	Geologist
Georg Neumayer	Scientist
Roderick Murchison	Director-General, British Geological Survey
Charles La Trobe	Lieutenant-governor of Victoria
Richard Owen	Superintendent, Natural History Department, British Museum
Adam Sedgwick	Professor of Geology, Cambridge
George Foord	Chemist

Introduction

In 1851 the newly proclaimed colony of Victoria, in Australia's south-eastern corner, enjoyed great good fortune; viable gold deposits were discovered within its borders in this same year as its separation from New South Wales. Gold fever immediately gripped town and country alike. From the colonies and settlements of Australia, and across the world, thousands of men and women rushed to the rich fields north-west of Melbourne to seek their fortune. The land to Melbourne's south-east held a different promise. Damp and swampy, cut by inconsistent creeks draining to Western Port's bay, it lacked the glamour of boomtown areas Ballarat and Bendigo. No gold was to be found here – but this earth stored samples of another precious metal.

In 1854 European settlers identified fragments of iron, at first considered natural outcrops but later identified as meteoric in origin. Subsequent discoveries revealed a fall pattern in the form of a lengthy strewn field. These ferrous deposits, large and small, littered a mile-wide ribbon from Beaconsfield in the north to Langwarrin in the south, with the foremost specimens found close to the rural village of Cranbourne.

Among the plains of this area, known as Mar-ne-bek to its indigenous inhabitants, the highest prominence was a sand hill named Towbeet, revered by the local Boonwurrung for its spiritual significance and from where game and campfires could be easily identified in the adjacent landscape. Today, mining activity has reduced Towbeet to a low mound, and picturesque botanic gardens overlay the site. I'm here early on a bright morning in March, the weather is mild but warming and the sky a hazy backwash patterned with a damask of plump cumulus. The park is a marvellous assemblage of walks and waterways, bushland, and immersive displays, and I spend several contented hours among its collections before my curiosity draws me to the exit. Because I'm not here just for the gardens, and I've enjoyed enough the pleasant anticipation of carrying out the real reason for my visit. Outside the gates I take an uphill walk via gravel track through native grass and low bush to the summit of nearby Trig Hill. A short climb on a wooden staircase brings me to a wide, circular viewing platform. A panorama opens up. The sweep of surrounding terrain can be appraised, and I get a sense of the geographic arena in which the cosmic visitation occurred.

To the south, flat country meets the head of Western Port and envelops the moated French Island with a fringe of estuarine mangrove. The dusky You Yangs wrinkle the horizon in the west, over the pearly grey waters of Port Phillip, and the distant Brisbane Ranges swell up in undulations to meet Mount Macedon's crown. Around to the north the brooding hump of the Dandenongs stands dominant above forests of box ironbark, then proceeds leisurely eastward into the Dividing Range's layered blue folds. It was from this north-eastern bearing that the meteor made its hurtling entry through the atmosphere; perhaps at night, probably emblazoned with a roiling tail, almost certainly with an accompanying air burst that would have lit the country with a flash visible from Cape Otway to Cape Howe.

I shade my eyes and imagine such a rent in the sky, as up high a flock of ibis dip and flow in train. Much of the meteor's body disintegrated in the intense heat and violence of its fiery crossing. But its size and entry trajectory meant enough mass remained intact to fall to Earth as a line of meteorites, fashioning a remnant celestial brooch whose central stone was set in Towbeet's swampy soil. Its recognition by colonists would lead to a drawn out and sometimes rancorous exchange. On one side, colonial scientists keen to retain local artefacts. On the other, a home institution that pursued a metropolitan agenda and assumed itself the superior location for placement of what would become, famously, the Cranbourne meteorite.

1
A Memo of all the Facts

As if rock didn't survive,
and dust didn't dance on air.
Jill Jones 'When Planets Softly Collide'

In the western desert of Australia, the Jack Hills are a belt of folded rocks that contain the oldest material on Earth: crystals of zircon formed over four billion years ago. Further west, on the continent's edge at Shark Bay, is a briny colony of living fossils — marine stromatolites — the world's most ancient life forms. In South Australia the Flinders Ranges' stony pleats sequester pre-Cambrian imprints of soft-tissue, multicellular organisms thought the earliest examples of such: the Ediacaran biota. The prodigious span of time is firmly branded on the crust of the great south land, and those zircon crystals connect our planet's thin coating, and us, with a past incredibly deep and distant.

More than four and a half billion years ago, on an arm of the Milky Way galaxy, the Solar System began to form from a nebula of dust and gas, the debris of ancient supernova cataclysms. This vast cosmic litter, known as a molecular cloud and light years across, was a gruel of hydrogen and helium, speckled with heavier elements. It gradually cooled and eventually collapsed under the effect of its own gravity. Then it rotated and flattened, spinning faster and aggregating material at its centre to cleave a hot, dense proto-star. Within the thinning disk, gravitational attraction and heat summoned gas and dust particles into molten droplets of matter, forming tiny spheroidal grains called chondrules. These accreted into bodies of increasing size, moulding the rocky inner planets, capable of withstanding the heat of their proximity to the new star.

With scarce metallic source material in the parental solar cloud, the heavy inner planets remained relatively small. Not so the icy outer bodies. They too were fashioned from gas and dust, but also captured the lighter hydrogen and helium swept from the system's centre to its cooler margins by an early solar wind. These cold worlds were giants in comparison to their rocky neighbours: Uranus and Neptune accruing huge solid volumes of otherwise volatile components such as ammonia and methane, and the enormous Jupiter and Saturn developing over 100 satellites, even while orbiting their own parent star. The outer giants are so large they comprise 99% of the total mass circling the Sun.

Victoria's first year was also its most tumultuous. It began with a terrible conflagration. After weeks of isolated fires up country, and a portentous annular eclipse on the first day of the month, on 6 February a bushfire of incredible ferocity exploded across Port Phillip. Arriving after an intense drought period in 1850 and fed by high temperatures and roaring northerly winds, a phalanx of blazes swept down from the Plenty Ranges and set the district alight. Twelve million acres were burned in simultaneous flare-ups from Portland to Western Port. One million sheep, and many thousands of cattle, were killed. A dozen people died, including the piteous loss of a shepherd's wife and her five children at Diamond Creek. At one point Melbourne was under threat, with much black smoke and a rain of embers converging on worried occupants, but a southerly wind change and drop in temperature around sunset spared the town. English author William Howitt, writing later, noted with some hyperbole 'a whole country of 300 miles in extent, and at least 150 in breadth, was reduced to a desert.' In an echo of Pliny the Younger before Vesuvius, he described the startling atmospheric effects on that day. '… While far out at sea, there were driven clouds of dust and ashes, which covered the decks of ships like snow, and obscured the midday sun.'[1] So intense and widespread was the inferno that thick smoke was reported across Bass Strait in northern Tasmania.

The essence of the day's confusion and fear is captured in William Strutt's confronting painting 'Black Thursday,' a dramatic composition showing settlers, livestock, and wildlife fleeing en masse before an enveloping cloud of smoke and soot. Although rendered in the European tradition, with the frightened mob resembling a grand cavalry charge toward the viewer, it projects a distinctly Australian tapestry: destructive natural forces testing human endurance.

The next twelve months would bring a marked reversal in Victoria's fortunes. In the week after Black Thursday, on a creek north-west of Bathurst in New South Wales, a group including Hampshireman Edward Hargraves made Australia's first significant gold find. When this was proclaimed in mid-May it began a rush to the new diggings around Ophir – Hargraves' name for the fields adjoining his lucky strike – and an exodus from Australian settlements, including Melbourne. Alarmed at the loss of population to Bathurst and the likelihood of economic downturn, a group of Melbourne councillors and prominent citizens convened a 'Gold Discovery Committee' in early June, meeting at the Mechanics Institute in Collins Street and proposing a resolution:

> *That this meeting is of opinion that gold in considerable quantities exists in close proximity to Melbourne; and that a subscription ought to be forthwith entered into for offering a reward to any person or persons who shall disclose to a committee to be appointed, a gold mine or deposit, capable of being profitably worked within a distance of 200 miles of the city.*[2]

They did not have long to wait. By a quirk of geology, a liberal seasoning of alluvial gold sprinkled patches of Victoria's crust, and small amounts had already been uncovered as early as 1849. James Esmond made the first serviceable find near Clunes, north of Ballarat, in late June 1851. The Irishman had returned to Australia in company with fellow 'forty-niner,' Edward Hargraves, after both laboured without success on the California gold fields. Almost simultaneously, other finds were announced at Warrandyte and again near Clunes, then a strike at Bunninyong in August. In September Mount Alexander gave its name to a nearby field of fabulous riches, from which four million ounces of alluvial gold would be extracted in the next two years. More discoveries were made at Bendigo in December, confirming Melbourne's north-western periphery as among the richest gold-bearing provinces of the world. In 1852, thousands of gold seekers going through the Black Forest to the fields of Castlemaine and Bendigo were surprised to see the trees putting forth a fresh display of green leaves amid the blackened remnants of the previous year's fires.[3]

Victoria's Lieutenant-Governor Charles La Trobe presided over the golden events of the early 1850s, and the almost immediate helter-skelter influx of new arrivals with its myriad attendant problems tried him to the point of exhaustion. Referring to a preliminary gold sample he had seen in 1849, he confided to a friend 'The truth is the discovery of a good vein of coal would give me more satisfaction.' Events of the first years of the rush, which he called 'the gold outbreak,' did not change his opinion. The young colony was in turmoil, and the mass departure of people from Melbourne to the goldfields meant a shortage of police, tradespeople, labourers, and even civil servants.[4] With an elected, and truculent, Legislative Council bristling at the appointed La Trobe's authority over potential gold revenues, he imposed a gold licence in an attempt to curb the rush of diggers. After numerous disputes the Colonial Office intervened to allay the vocal concerns of miners and mining interests, but by year's end La Trobe had had enough. Weary from the aspirations of the district, and now colony, being freighted onto his shoulders his resignation noted 'that constant strain upon the mind more than the body, which the weight and character of my public duty, particularly of late, have brought with them.'

Victoria's population trebled in the three years from 1851. The gold fever brought dozens of shanty-town mining communities to the new colony's rivers and creeks. Some, like that on the Yarrowee River, later Ballarat, would thrive and become major cities. Others declined almost as rapidly as they arose. After these waterways gave up their alluvial treasure, gold mining would take on a more industrial manner as deep leads of quartz were scrutinised for threads of the precious yellow metal. Melbourne, perched above the Yarra at the head of Port Phillip's broad bay, was the entrepot through which immigrants and golden wealth flowed. Despite the ructions of the early rush it would prosper — beyond the dreams of avarice.

* * *

The Melbourne Exhibition of 1854 was the first exposition organised in Victoria in the nineteenth century. The colony was in only its third year, but the recent flood of gold seekers had swelled Melbourne's population to 76,000, from census figures of 29,000 three years earlier, and the former Port Phillip District was home to more than 220,000 people. Civic and cultural foundations flourished. Construction began on two great institutions for the furtherance of knowledge and communication in this year: the University of Melbourne and the State Library of Victoria, and The Age newspaper was first issued.

Commissioned by Queen Victoria, the exhibition was intended to attract displays worthy of contention for the Paris Exposition Universelle, to be held the following year. A purpose-built structure was erected on the corner of William and Little Lonsdale streets, designed by architect Samuel Merrett. Melbourne's leading daily newspaper, the *Argus*, reported it 'a very handsome and spacious palace,' and 'a *tout ensemble* of elegance and utility combined.' Its iron-and-timber frame supported nearly 200 windows and a glass roof, and was a reflection in miniature of Hyde Park's Crystal Palace.

The exhibition's official catalogue is a revealing window to Victoria's past, its cover page decorated with a stanza from Scottish Romantic poet Thomas Campbell's *The Pleasures of Hope*. The name of Redmond Barry, acting chief justice and later to send bushranger and fellow Irishman Ned Kelly to the gallows, glares down from atop the list of the exhibition's commissioners. He and his fellow notables precede the exhibition ode (10 stanzas in qualitative metre), a map of the presentation hall, and finally a list of the displays themselves.

And what a trove of curiosities they are: steam engines and electro-magnetic machines, dessert knives and leather goods, boots with revolving heels, native wood, and stuffed animals. The display of William Morton, from Loddon, was a peculiar combination: a 'Lady's private bath' and an 'Improved plan for operating on cattle without roping them.' Crikey.

Then on page 10, among the rock specimen section, is a nondescript entry listed as item 16:

> *Scott, James A., 32 Little Collins-st. W., Farrier. — Specimen of Iron from Western Port, and a horse-shoe made from it.*[5]

Some iron and a horseshoe. Hardly the stuff of colonial presentations, let alone grand Parisian exhibitions, one might think. But Scott's horseshoe was unusual, although more in composition than appearance. Iron in its elemental form is vanishingly rare in Earth's crust. Most sank to the planet's centre during the differentiation phase of early formation, billions of years ago, when the interior started separating into the crust-mantle-core layers we recognise today. Usually, crustal iron is found in its oxidised profile, as iron ore, of which the compounds magnetite and hematite are most common. The smelting of iron ore in high-temperature furnaces to separate its constituent elements has been carried out for several thousand years. Blacksmiths, and farriers like Scott, used charcoal and bellows to generate a fire of suitable temperature — not hot enough to melt the extracted iron, but sufficient to produce a workable mass, or bloom. This could be hammered into tools and ornaments, or indeed horseshoes. Pure iron, when found in the crust, is almost exclusively confined to meteorites. So, while Scott's sample was sourced from the Western Port district, it was not native to the area, as thought at the time. In fact, it was not even terrestrial in origin – it had fallen from the stars.

In October 1859 a conference of a different nature was held in Melbourne. It was convened to discuss the viability of an ambitious undertaking – the extension of the railway line from Melbourne, at Brighton, to the coal fields at Cape Paterson in south Gippsland, past Western Port. Decades earlier, explorer William Hovell and settler Samuel Anderson found small deposits of the mineral in the area, and these had come in for more recent examination. In 1854 the Government Geological Surveyor, Alfred Selwyn, reported to Parliament on the cape's coal formations via a survey of the coast between Western Port and Anderson Inlet. He noted 'The carboniferous formation appears to occupy a very considerable area in this district.' But he also concluded that 'none of the seams seen on the coast are likely to provide sufficiently permanent in thickness or extent, to pay for the great outlay required for the construction of a tramway fifteen miles in length, and other works indispensably necessary before any large supply of coal could be brought to market.' More drilling of sample bores by a government contractor in 1858 did not change Selwyn's view. Although he adjudged Cape Paterson coal superior to that of New South Wales, he was still cautious on the extent of deposits, and recommended more exploratory shafts be sunk.

Despite these underwhelming mineralogical findings, the conference was well attended. Delegates included Melbourne Town Clerk Edmund FitzGibbon, the surveyor general, and the chairmen of municipalities Brighton, Prahran, and Dandenong. Interest came from further afield, in particular the colonists of Gippsland, east of Cape Paterson. This despite the *Gippsland Guardian's* editorial in September which opined that the proposed link, while 'valuable because of the ready means of intercourse which the railway would offer' would 'not be commercially viable to private operators.'[6] The auriferous riches of Victoria's 1850s boom years, in terms of investment and population growth, did not extend to the colony's near-eastern localities. The cadastral map of early Victoria has the Western Port area falling within the county of Mornington, which the 1854 census reported as having a grand total of 1,372 persons, male and female, residing within its various parishes. A railway to, or through, this area was bound to stimulate commerce and provide additional benefits to the existing population.

Among the discussions on potential line routing, funding options, and levels of government participation, one Alexander Cameron put forward a novel submission. Another of many Scottish Presbyterians settled in Western Port, Cameron was a well-known pastoralist and founding trustee of the Scots Church. He held the lease on the Mayune run, through which the track to south Gippsland passed, and around which the small community of Cranbourne began to coalesce in the 1840s. A township reserve was set aside there in 1852 (from parts of the Mayune and Towbeet runs) and blocks were allocated in the government survey of 1856, which included a Cameron Street among its subdivisions.

The squatter was plumping for the mooted railway, and for Cranbourne to be included in any proposed route. He exhibited to the conference delegates samples of local iron, sourced from Cranbourne's environs. Geologically, the region was comprised of Silurian sandstone, but Cameron believed the area included an extensive iron deposit and wanted to promote its commercial possibilities in support of his proposal.[7] His example irons included the horseshoe last seen in the 1854 exhibition, and a fist-sized lump that was most likely its earlier accompaniment.

The shoe was forged in the smithy of Cranbourne blacksmith James Nelson, himself to later craft the ironwork on the first Princes' Bridge across the Yarra.[8] The source of its metal was a large mass that lay on the property of Hugh McKay, 'Murrangang,' in the parish of Sherwood, a few miles south of the allotments of Cranbourne. In the settler narrative, a horseman tethered his mount to what he thought was a protruding tree stump but, surprised at its metallic texture, discerned its true nature.[9] An 1857 Surveyor General's map describes the locale as 'sandy hillocks covered with heath.' It entitles numerous blocks in the vicinity with the name John Bakewell, a Nottingham-born absentee landlord who had split his large run, Tooradin, into several parcels, including one named after the famous forest of his home county. Like Cameron, Bakewell had a street named after him in the new Cranbourne grid. In a map compiled the following year Section No. 39 is shown as McKay's, held under pre-emptive right. It is bordered to the east and north by several allotments under the ownership of James Bruce, a Scotsman who had made his purchases in 1858. The country was labelled as 'medium agricultural soil thickly timbered with gum, honeysuckle and she oak.'

Edmund FitzGibbon visited Cranbourne in early 1860 to investigate Cameron's claim.[10] The Cork-born town clerk claimed descent from a Hiberno-Norman hereditary knighthood, the White Knight, and lived in London from an early age. Educated privately, he worked for a committee of the Privy Council before the lure of gold brought him to Victoria in 1852. After a year on the Castlemaine diggings he took a role in Melbourne, proof-reading papers in the new Legislative Council chamber. By 1854 he was working in a clerical role at the Melbourne city council, where he was gazetted town clerk in 1856 upon the resignation and death, respectively, of the position's two previous occupants. Although trained as a barrister, FitzGibbon would have had some knowledge of metallurgy from his time on the goldfields, and when he viewed the outcrop in McKay's paddock he recognised it as a lithological interloper. While there were areas of ferruginous sandstone to the immediate east, this was something else again – and comprised entirely of iron.

After FitzGibbon's Cranbourne foray the governor of Victoria, Sir Henry Barkly, had something catch his attention when on government business at the town hall in Melbourne. He 'saw on the mantel piece of the Town Clerk's room a horseshoe with an inscription stating that it was forged from a lump of virgin iron cropping out from the ground near Cranbourne.'[11] Barkly made inquiries about the horseshoe's provenance, no doubt questioning FitzGibbon as to his own impressions. What he learned further piqued his attention and he invited FitzGibbon to 'draw up a memo of all the facts' and present these to the Royal Society of Victoria, of which he was then president.

Barkly took a keen interest in scientific matters. Appointed governor in 1856, he was the son of a Scottish merchant with ties to the West Indies. Initially in business and then a member of the House of Commons for three years from 1845, he was placed in gubernatorial roles in British Guiana in 1848 and Jamaica in 1853. The Colonial Office looked favourably upon his Caribbean success, and it led to the Victorian appointment. He supported intellectual pursuits in the colony and was a co-founder, with other local notables, of the Royal Society.

The society was the centre of scientific conversation and promotion in the fledgling colony. Modelled on its prestigious London forebear, it had its beginnings in two recently founded local organisations. In 1855 the Philosophical Institute of Victoria formed as an amalgamation of two learned societies established in the previous year, the Philosophical Society of Victoria and the Victorian Institute for the Advancement of Science. Dedicated to the gentlemanly pursuit of scientific endeavour, the new entity included the only four scientists in the colony with requisite training and adequate experience to accompany the title. The German botanist Ferdinand Mueller and the geologist Alfred Selwyn had arrived in 1852, and appointees to University of Melbourne, professors William Wilson and Frederick McCoy, took up their positions in early 1855.[12]

Other members represented the wealthy and influential of Melbourne society, including the recently appointed Barkly, Chief Justice Sir William

Stawell, and Lord Mayor Dr Richard Eades. Complete with executive officers and a 12-member council, this combination of scientific professionals and enthusiastic amateurs conducted proceedings on a regular basis. These usually involved the reading of technical papers on topics ranging from the commonplace, 'On a New System of Ventilation,' to the arcane, 'On the Nidification of the Coach-whip Bird and White-fronted Epthianura.' Proceedings were painstakingly recorded and published as annual volumes of 'transactions,' including, in the early years, significant papers by Mueller on new species of flora found in the colony. In 1859 royal assent was granted for a change of name to the Royal Society of Victoria.

Architects Reed and Barnes were commissioned to design a building that would reflect the status of the newly chartered body and provide a space for practical application of its various activities. A government grant secured a wedge of land on the north-western edge of Melbourne's grid, bordered by Exhibition, La Trobe, and Victoria streets, and in 1859 the initial structure was completed. In the Renaissance Revival style and finished in red brick, it comprised three northern bays and a vaulted meeting hall, with Tuscan pilasters, arched windows and rounded corners as distinctive external features. The building's square-plan form remains in place today; heritage listed and tidily kept, it is a low-rise island of scientific serenity among Melbourne's steel-and-glass clutter. One of the last Philosophical Institute meetings was held there on 21 December 1859, the eve of the summer solstice. President Ferdinand Mueller's closing address suggested the symbols of 'concord and progress' to fulfil members' destiny and exhorted them: 'May the Tempest of discord never re-echo from these walls!' A worthy ambition, but one more noticeable in the breach than in the observance, and over the next few years the Cranbourne irons would see the Royal Society rooms host much discussion, debate, and dispute.

2
Two Masses of Malleable Iron

The meteor was a dove of amethyst
that hatched from an ancient storm.
M.T.C. Cronin 'It was in Venezuela that light was forged'

Not all the building material of the Solar System was converted to planets. In the late 18th century German astronomers Johann Titius and Johann Bode hypothesised that planets occupied predictable trajectories, measured in distance from the Sun. The Titius-Bode 'law' – more a rough rule – related the mean planetary distances to a simple progression of numbers. This mathematical expression held up for the inner planets, and the known gaseous giants of the time, Jupiter and Saturn. Uranus's discovery in 1781 also fit the postulation.

But the sequence had a gap in the cold space between the orbits of Mars and Jupiter, the known fourth and fifth planets respectively. Bode predicted the existence of a body there, and in early 1801 the Italian priest and mathematician Giuseppe Piazzi obliged, observing 'slow and uniform' movement in 'something better than a comet' and discovered the asteroid Ceres, so satisfying the formula. Ceres was not alone, however. It was only the first object identified in a large field populated by millions of bodies and occupying this fifth 'slot' from the Sun. Taking their name from the Greek *asteroeidis* for their star-like appearance, these twinkling entities range in size from dust particles to monoliths many kilometres in diameter. In combination, they form the Asteroid Belt.

In its primeval state the Asteroid Belt was a rocky solar necklace with a total mass about equivalent to that of Earth. But massive, marbled Jupiter's immense infusion of orbital energy prevented the planetesimals of the belt from accreting into larger planets. Washed by the periodic resonances of

> their Jovian neighbour, these pent-up wrecking balls collided with such frequency during the Solar System's early history that they repeatedly shattered, ejecting fragments from the belt and depleting its overall mass until what remains today is equivalent to only 4% of our Moon's mass. Over half this amount is tied up in just four of the largest bodies: Ceres, Vesta, Pallas, and Hygiea. Millions of smaller bodies accompany them.
>
> Most planets orbit the Sun on a plane extending from their star's equator, called the ecliptic, and the majority of asteroids also do so. Even so, many asteroids have a tilt in their orbital inclination, up to approximately four degrees. The belt is really more of a great speckled doughnut in space.

In line with Henry Barkly's wishes, on 4 June 1860 Edmund FitzGibbon attended the scheduled ordinary meeting at the Royal Society's rooms and 'exhibited specimens obtained from two masses of malleable iron, lying in the vicinity of Cranbourne.' His description of the McKay sample is the first recorded in detail:

> *A mass lying on the land of a Mr. McKay, on Section 39, parish of Sherwood, distant about three and a half miles in a southerly direction from the township of Cranbourne. It presents a tabular face nearly level with the surface of the land, and somewhat of a triangular shape, the edges measuring respectively about 31, 33, and 38 inches. A trench excavated around it has revealed its sides to an average depth of about 30 inches, the bulk of the mass becoming greater as the depth increases, inducing a belief that the weight of the portion visible amounts to about four tons.*[1]

Whilst in Cranbourne FitzGibbon heard of another native iron example in the district and visited the site where it lay. He also described it to the society:

> *A mass similarly bedded in land belonging to a Mr. Laneham, section 39, parish of Cranbourne, distant about two miles eastward from the township, and about four miles north-eastward from the mass just described, similar to it in its general characteristics, but apparently not more than one half of its bulk.*[2]

FitzGibbon presented the members with the horseshoe and another fragment of a much smaller third specimen, about seven pounds in weight, given to him by Hugh McKay. This piece was apparently half of a block, found within a half-mile of the McKay mass, and used for some years as an andiron in the hearth of a local labourer's house before being broken in two and the other part lost. FitzGibbon told the assembly he acted upon Governor Barkly's suggestion 'with a view of inducing inquiry and experiment for the purpose of deciding whether the deposits are native iron, and bear affinity to the local formation.' But he had now formed that view that these irons, and the masses from which they derived, were meteoric in nature 'from their general resemblance to aerolites known to have fallen in other parts of the world, and which have been carefully analysed and described.'

FitzGibbon did not reference sources in support of his findings but he was quite correct in his deduction, even if the quaint term 'aerolite,' based on Greek roots for 'air' and 'stone' has fallen into disuse. He told of the larger specimen that 'the upper surface is studded with apparently oxidised blisters, which are easily detached in scales, and which in some instances contain a non-magnetic metallic substance approaching to the character of black lead.' These characteristics align with other iron meteorite finds to that date and describe the rusty looking oxyhydrated mineral lawrencite, and the dark fusion crust formed on the outside of the body under the high temperature and pressure of atmospheric entry.[3]

By convention, meteorite finds are named according to their nearest locality. Multiple related finds in a common area, called a strewn field, each take that name and have a numeric suffix for separation. The three iron specimens, in order of size and likely date of discovery, would be referred to thereafter as Cranbourne No. 1, 2, and 3. Their irregular shapes challenge easy comparisons, but No. 1 is roughly equal to a club chair in its volume, while No. 2 is closer to a footstool in size. To grasp their staggering weight, attributable to a very high iron content, imagine the above furniture being carved from a blacksmith's oversized anvil.

As the owners of the land on which the meteorites were found, McKay and James Lineham (incorrectly referred to as 'Laneham' by FitzGibbon) were their presumed owners under principles of English common law, and had been in effective possession since taking up their allotments. Although British settlers disported themselves in an arena of great antiquity there was rarely any acknowledgement of such, and no thought as to the area's Aboriginal inhabitants being potential custodians appears to have been entertained. The settlers brought their own discovery narrative, emboldened by the exploits of Hamilton Hume, Thomas Mitchell and Pawel Strzelecki, who struck out through 'pristine' wilderness and opened up stock routes and new grazing lands. That the kangaroo and emu had long bounded over the land and taken sustenance from its native grasses mattered not. Indigenous voices were trammelled to docility.

Hugh McKay had sailed from Greenock, Scotland, in December 1839 on the *Glen Huntly*, a vessel chartered under the Bounty Scheme to assist immigration to Port Phillip. A grounding soon after its initial departure from Oban might have alerted the superstitious that this maiden voyage would be an unlucky one, concerns heightened when scarlet fever, measles, and some smallpox cases were reported aboard. By the time it cleared the Firth of Clyde seven young lives had been lost. And indeed, when the barque arrived in Port Phillip the following April it was flying the yellow flag that indicated fever aboard. Ten of the 137 passengers had died on the journey, most likely from typhoid. Another three would succumb in the following days at the quarantine station that then-Superintendent Charles La Trobe hastily convened at Little Red Bluff, later Point Ormond. Also on the ship was Jennet Smith, a 23-year-old from Ardclach travelling with her sister, Robina, and other family members. It appears a romance blossomed between Jennet and Hugh, ten years her senior; maybe on board, possibly in the uncertain days after their disembarkment. They married by year's end.

James Lineham was a Bedfordshire immigrant, born in 1824. In 1848 he married 20-year-old Charlotte Ridgway of Buckingham and took passage for Australia on the *Francis Ridley*, arriving in Melbourne during the heat of February 1849 with his pregnant wife and two hundred other agricultural labourers. After a few years at Keilor the young family made for the Ballarat gold fields. Two more children followed, and James collected enough on the diggings to purchase the property east of Cranbourne in 1854. He built a wattle-and-daub house with thatched roof to shelter a growing brood, 12 children in all by 1873. The work was hard; trees were felled for fencing and oats were hand-sewn and cut with scythes. Charlotte washed clothes at the creek's edge or carted water up from streams and gullies. Fetching supplies meant a day-long bullock ride to Cranbourne and back. Mass on Sunday was a brief respite from labouring, although 'Shanks's pony' was deployed for transport, and New Year's Day formed the only holiday – a trip by horse and dray to Tooradin for a picnic, sure to please the children. The family would remain in the area for generations.

At the Royal Society active discussion ensued among members on FitzGibbon's exhibits and other specimens presented at the June meeting. Frederick McCoy displayed samples of European meteoric iron. He would play, with Ferdinand Mueller — also in attendance — a leading part in subsequent events. What FitzGibbon did not disclose, nor see entered into the meeting's minutes, were several discussions he had with the Cranbourne landowners. As well as McKay's presentation of the moiety of No. 3, FitzGibbon would later write that the farmer had offered him No. 1 in its entirety, 'in free gift.'[4] Furthermore, Lineham had offered him No. 2 for five pounds. In an early turning point of what would become a multi-year tussle, both offers were declined. FitzGibbon's grounds for doing so seem naïve, given future proceedings. 'I thought the Government should own them for our Museum; and with that object, drew public attention to their existence and position.' His note certainly drew attention in Europe. Emperor Franz Joseph of Austria made inquiries through Barkly, and the governor's reply to the monarch, conducted via Mueller, included the presentation of a small sample taken from No. 1.

* * *

It was a busy period for the Royal Society, which included the preparation and despatch of the Victorian Exploring Expedition – Robert Burke's and William Wills' famous and ill-fated crossing of the continent – for which it was the primary sponsor and organising body. The expedition launched from Royal Park on 20 August 1860. The build-up and thrilling, although somewhat disorderly, departure of their grand exploratory party meant the

Royal Society let the Cranbourne meteorites recede from the collective memory across the spring and early summer of 1860. Despite the interest aroused by FitzGibbon's address and paper, eight months elapsed before a society representative visited Cranbourne and sighted the specimens. This was a young magnetician and hydrographer, Georg Balthasar Neumayer.

Born in south-western Germany in 1836, Neumayer's inquiring mind engendered an early attraction to the sciences and he studied geophysics and hydrography at Munich University. Polar exploration, terrestrial magnetism, and navigation were other areas of interest. Travelling to South America and Australia in turn, and paying for his passage with work on board ship, he mined for gold at Bendigo and performed research at the Rossbank Magnetic Observatory in Hobart. Excited by the southern continent's scientific possibilities, in 1854 he returned to Germany and set about raising money and support for his plan to establish an observatory in Melbourne. It was soon forthcoming. He secured backing from the great geographer and explorer Alexander von Humboldt, and King Maximilian of Bavaria provided substantial funds for instruments and equipment.

Arriving in Melbourne in early 1857, Neumayer was initially denied a location in the Botanic Gardens when dissenting voices questioned the usefulness of purely magnetic observations and his proposed magnetic survey of the colony. Beyond navigation, the potential utility of knowledge of Earth's magnetic field was not well understood in the mid-19th century. But the closure of Rossbank in 1854 brought about the need for a continued observation of this data in southern hemisphere locations, to tie in with other stations operating around the globe.

In the end, the inclusion of weather and oceanographic data collection in Neumayer's proposal meant he was permitted to conduct observations on Flagstaff Hill. By May 1858, purpose-built rooms were operational, with magnetic observations taken on the hour. William Wills was employed there as an assistant. Neumayer also traced the paths of meteors from the new building using a clever instrument of his own invention. The meteorograph was a sighting device that used a plate on a telescope's equatorial mount to record the start and end celestial coordinates of individual meteor tracks. The efficacy of the device was only surpassed years later, with the advent of astronomical photography.[5] An 1858 lithograph by Exeter artist George Rowe, who worked on Victorian diggings for five years, shows a rare panorama of the early Melbourne township from Flagstaff's northern aspect. The garden's observatory infrastructure is in foreground, with crinolined matrons holding parasols in conversation nearby, and goldfield-bound diggers making off west in the distance.

Neumayer's field work began later that year. He travelled tirelessly, in often difficult and sometimes dangerous conditions. In the six-year duration of his Victorian tours, collecting data and camping out, he made observations at over 230 locations, covering 11,000 miles on horseback, by cart, and on foot. His survey findings eventually made their way into print in 1869, after his return to Germany, as *Results of the Magnetic Survey of the Colony of Victoria—1858–1864.* It was a demonstration of the colony's golden fame that such a bright scientific mind was drawn into its orbit. But while Melbourne enjoyed an enviable reputation among the cities of the world by the mid-1850s, the Port Phillip District's beginnings were characterised by fits and starts, mis-steps, and mis-calculations.

> The early history of Victoria was entwined with the sea. Long lapped by the tides of the Southern Ocean, and with a shoreline serrated by ice age glaciers, its land awaited the visitations of ocean-borne navigators. British captain James Cook, aboard *Endeavour,* was the first European to sight land, at Point Hicks, on the first of his great voyages in 1770. Next was Royal Navy surgeon-turned-explorer George Bass in 1797, piloting an open whaleboat south from the now-established penal colony at Port Jackson. He reached as far as Western Port, exploring Phillip Island and surrounds, and returned the next year in company with Matthew Flinders aboard the sloop *Norfolk,* looking for confirmation of a strait between Van Diemen's Land and the northern land mass.
>
> More Royal Navy men followed. Lieutenant James Grant sailed from Portsmouth in early 1800 in charge of the *Lady Nelson,* and by December had travelled Bass Strait from west to east, naming Portland Bay, Cape Otway and Cape Schanck, noting but passing by the entrance to Port Phillip. Only weeks behind him was the precocious John Black, en route from Cape Town to Sydney town in the brig *Harbinger* with a cargo

of wine and rum for the thirsty colony. Grant later mapped the shore between Wilsons Promontory and Western Port. The land slowly revealed itself.

In Sydney, Governor King commissioned Grant's First Mate, Acting Lieutenant John Murray, to continue the south coast exploration in November 1801. It was Murray's own First Mate, William Bowen, who was the first Briton to enter Port Phillip, reporting it 'a great and noble sheet of water.' Murray's group stayed for almost four weeks, and he noted the first interactions with the indigenous inhabitants there. These were initially amicable but thereafter violent – the first of many such episodes.

The warring Britain and France now sent two famous navigators, Matthew Flinders and Nicolas Baudin, on exploratory missions to the south land. Flinders entered Port Phillip only weeks behind Murray, climbing local landmarks and enjoying more peaceful encounters with the local clans. Baudin mapped the east coast of Tasmania before heading north to Western Port. Later he had an unexpected seaborne encounter, in what are now South Australian waters, with the young English navigator whose very chart he had been referencing through the strait, Matthew Flinders. Baudin's companion ship, *Naturaliste*, was separated from its flagship and separately made Western Port in April 1802, where its company surveyed the waterway and named French Island.

From the British base in Sydney harbour, hemmed in by the as-yet impassable mountainous terrain to the west, and aware of the French incursions in the south, Governor King sent a surveying party to Port Phillip, but only desultory reports followed their excursion. From Britain, a concerted attempt at settlement was made in 1803 under Lieutenant-Governor David Collins. However, this lasted only six months in the challenging country of Sullivan Bay to the south of the port area before Collins packed up for Van Diemen's Land and founded Hobart. And with that, seven years of intensive inspection of Port Phillip and Western Port by European navigators were concluded. These explorers were largely tied to their vessels, and only skirted the edges of the landmass, ever careful to replenish ships' stores, repair hulls, and record soundings. Their sorties ashore were usually in search of game or water, or to scale a hill to enhance their sea-level view of the country. It was not until 'overlanders' ventured south from Port Jackson that an appreciation of the interior of the future colony was gained.

In the meantime, a veil was drawn across the land, and the Victorian coastline receded from history's view; its shores and islands plumed by sea spray, its hinterland lashed by winter storms and baked brown by the summer sun. The great bays of Port Phillip and Western Port settled into their tidal rhythms, and wind and rain gradually erased the remnants of Collins' short-lived industry at Sullivan Bay. The land returned to its habits of millennia.

It would be a brief hiatus. In 1824 a young Australian pathfinder named Hamilton Hume was partnered with William Hovell, a Norfolk-born ex-Royal Navy man, late of the Hawkesbury and 11 years his senior, to explore the colony's southern reaches. Despite Hovell's subsequent petulant interference, Hume's bushcraft and leadership brought the party onward, encountering the Murray River, opening up good grazing tracts and farmland, and charting passage through the low pass in the Dividing Range at present-day Kilmore. They made the southern coast near Corio Bay, Hovell mistaking the location for the more easterly Western Port, before turning north and returning home in half the time of their outbound journey.

For a low-lying inlet with minimal anchorage and poor soil, Western Port attracted attention beyond its apparent virtues. Another French navigator would soon sail into its shallow waters. In November 1826 the grandly named Jules Sebastien Cesar Dumont d'Urville passed through Bass Strait in the corvette *Astrolabe*, during his second circumnavigation. After completing a survey and soundings around its contingent islands, he headed north and crossed, unseen, with ships from Sydney bound for Western Port. After years of inconclusive probing of the southern coast, and one abandoned settlement, it was again the prospect of French interlopers laying claim to territory on New South Wales' margins that spurred decisive action from the British colonial authorities.

On a smaller scale to the 1804 Port Phillip settlement under David Collins, in November 1826 Governor Darling despatched south a group of convicts, officers, and soldiers to set up a mini-colony on the eastern shores of Western Port. Erstwhile explorer William Hovell was in the party, and made amends for the Hume expedition with a series of excursions to the north and west, taking in the shores of Port Phillip. Hovell and his companions were the first Europeans to trudge the swampy plains north of Western Port. It was country of overlapping usage by the Boonwurrung and Gunai, with southerly running creeks, and lay under the Dandenongs' ridge line further north. A scattering of iron fragments nestled on that ground. Did Hovell perhaps skirt this strewn field, unknowingly, on his way across to Port Phillip? The penal settlement lasted little more than a year before it too ended in withdrawal. In the years to follow it would fall to private operators to arrive, and stay, in Port Phillip and fill the breach left by government inaction.

But another overland expedition, not intended to take in Victoria, would come first. In 1836 New South Wales Surveyor General Thomas Mitchell led an expedition down the Lachlan, Murrumbidgee, and Murray rivers before turning south, against instructions, and traversing the bountiful plains to Port Phillip's west. Mitchell was not one to miss an opportunity to self-promote, and had a prescient vision of what would become the American ideal of manifest destiny:

> As I stood, the first European intruder on the sublime solitude of these verdant plains, as yet untouched by flocks or herds; I felt conscious of being the harbinger of mighty changes; and that our steps would soon be followed by the men and the animals for which it seemed to have been prepared.[6]

Mitchell appraised a landscape which had indeed been prepared, but not for the purposes he proposed. Aboriginals had used fire to manage country for thousands of years, and nurtured a tapestry of forest, verge, and grassland in which to hunt game and harvest crops. Early explorers often marvelled at the seeming parklands they traversed, without consideration for the probable consequences of their expeditions. Mitchell's venture captured the imagination of the colony, and lent an exciting new title to the supine and fallow territory he revealed; 'I named this region Australia Felix, the better to distinguish it from the parched deserts of the interior country, where we had wandered so unprofitably, and so long.' His trailblazing journey, while the stimulus for an explosion of European expansion and settlement not seen in Sydney nor Van Diemen's Land, catalysed a drastic and irrevocable decline in Aboriginal welfare south of the Murray. Port Phillip's previously gradual unveiling would soon be overtaken by a land-grab of extraordinary proportions.

On 11 February 1861, Georg Neumayer travelled with a party to Cranbourne, then onward to Hugh McKay's property, at the invitation of Augustus Abel, a mineralogist and assayer of Ballarat. Burke and Wills had reached the Gulf of Carpentaria two days earlier. By evening the group was camped close to the meteorite. The next morning Neumayer surveyed the surrounds and measured No. 1 in detail, noting 'The earth around the meteorite had been removed to the depth of two feet; the lower part, however, was not visible, the hole being partly filled up with water.' Then he did something unusual.

Using a magnetised needle and silk thread, he approached the sides of the partially buried specimen, noting the northern polarity of its upper portion by the attraction effect on the southern end of the needle. Almost two feet down the meteorite the magnetism changed from north to south, leading Neumayer to deduce its height as approximately four feet. But the polarity change was not consistent, in distance from the top, as he investigated each side. He concluded 'that the shape of its lower part was that of a wedge; basing thereupon, I calculated the total weight of the mass to be 4.3 tons.'[7] This innovative method of calculating the dimensions and depth of the meteorite is the earliest identified occurrence of quantitative geophysical interpretation, being the calculation of dimension, volume and depth to source. Neumayer was ahead of his time.

Some small portions were taken from the specimen and measurements of specific gravity made. The young scientist conversed with the local settlers, reporting:

> *The mass itself was originally buried in the ground, a small piece only, 4 inches long, protruding above it and it was by this means that it was first discovered. I was told by some old colonists that they remembered the time when the natives used to dance around it, beating their serpentine tomahawks against it and apparently much pleased with the metallic sound thus produced.*[8]

The mention of 'serpentine tomahawks' is noteworthy. Hafted hatchets with ground-edge greenstone heads – serpentine being a form of greenstone – were in wide use among Aboriginal groups of south-eastern Australia. The stone heads of the 'tomahawks' noted here most likely came from the *Wil-im-ee Moor-ring* site (Mount William quarry), 48 miles north-west of the Yarra's mouth. These were heavily traded items in the commerce of Kulin peoples around Port Phillip. This passage is also a rare colonial reference to witnessed Aboriginal engagement with an object of astronomical significance. But regardless of any longstanding cultural, and possibly totemic, connection between local clans and the iron mass, while at McKay's property Neumayer was informed that No. 1 'was now in the possession of Mr Bruce, a gentleman residing on his farm near the locality.'

By the early afternoon of 12 February Neumayer and company were at Lineham's property, north-east of Cranbourne and 3.6 miles from No. 1's site, examining No. 2. This smaller body 'had been turned over on its broad side, rendering the entire mass visible.' Like the larger body, 'only a small piece of iron projected above the surface of the ground at the time when it was first discovered.' Neumayer approximated its weight to be 1.5 tons. He reports that 'Mr Abel bought this latter mass and made all the arrangements for having it brought to town.' Suzanne Lineham, daughter of James and Charlotte, reported in her later years that the Aborigines who came to the property 'worshipped' the fragment, and that it was so special they cried when it was taken away.[9] To her father it was no great loss, as he 'looked upon it rather as a nuisance and was glad to dispose of it.'[10]

McKay and Lineham were within their rights to trade their meteorites as they saw fit. So, between FitzGibbon's visit to Cranbourne – after the October 1859 railways conference but before his June 1860 Royal Society address – and Neumayer's February 1861 excursion, the specimens had changed hands. The town clerk seemed to have missed an opportunity. Who were these new owners?

3
Our Colonial Crystal Palace

Gliding is what I do,
Here at the finish, in the final hour.
It will be this way between the star clusters,
In the gulf between the galaxies.
Clive James 'The River in the Sky'

The bodies of the Asteroid Belt tend to clump into relatively dense zones of material and, like the rings of Saturn, lightly populated areas separate these. Called Kirkwood gaps, they are named after the American astronomer who first posited their existence and explained their origin. Once again Jupiter, which circles the sun once every 11.86 Earth years, plays a part. An asteroid whose orbital period is a simple fraction of Jupiter's is said to be in resonance with the planet, and repeatedly passes close to its Jovian oppressor in the same place in its orbit each time. But it doesn't stay in that orbit very long. Jupiter's gravity, compounded by the regular proximity, rips the asteroid out and away to other orbits.

So not all asteroids are found in the main belt. Some cross Earth's path or even reside inside our orbit: the near-Earth asteroids. Others are co-orbital with a planet, i.e. situated on the same track, and are perturbed by gravitational pressure from both the Sun and the planet to accumulate at an optimal 60 degrees ahead of, or behind, the larger body.

In Jupiter's case a Homeric nomenclature applies; originally those bodies positioned forward were called Greeks and those behind, Trojans, but all now take the latter name. Typically for Jupiter, it has by far the largest number of Trojans, over one million, in two separate swarms – outriders fore and aft in the gas giant's grand celestial motorcade. Many of these bodies take inclinations approaching 40 degrees to the plane of their primary's orbit, and form another speckled-doughnut track around the Sun, with mass equal to one-fifth that of the main belt. Earth has only one Trojan.

Mars' two moons are the tiny Phobos and Deimos, named for Greek mythological twins Fear and Terror, sons of the war god Ares. They were once thought to be captured asteroids, given their irregular shapes, inferred make-up, and observed radiation reflection. That was until their virtually circular orbits, behaviour not seen in captured bodies which usually have more erratic orbits, excluded that origin. The red planet's satellites are more likely to have accreted from the collision impact remains of a massive strike between Mars and an asteroid, a common occurrence in the early Solar System. Asteroid collisions' wreckage is a source of meteoroids, small remnant bodies of rock and iron hurtling through the Solar System, orbiting the Sun like their parent bodies.

The *Argus* was a Melbourne morning broadsheet founded in 1846, and which ran for over 100 years. Its edition of 9 September 1858 reported crown land sales for the month, conducted at the rooms of Tennent and Co. of Collins Street. Although a high proportion of the land on offer was described as 'sandy soil, covered with scrub,' and 'poor pasture land,' competition was brisk. The name of James Bruce was recorded as the new owner of several country lots listed under 'Sherwood – South of the parish of Cranbourne, at a distance of from 31 to 38 miles south-east from Melbourne. Parish of Sherwood, County of Mornington.'

Bruce hailed from Aberdeenshire, where Scotland's north-eastern lowlands elbow out into the North Sea and the Grampians rampart the western border. His birthplace is given as Old Deer, a small village on the River Ugie favoured with a monastery by the Irish missionary evangelist, Columba, in the sixth century. James' father Alexander farmed the calcareous slopes of nearby Mill Hill and with his wife Mary tended to their brood of five – James was the fourth-born and first of two boys. Alexander was of sufficient means to establish his sons on holdings in the district. As a young man in the early 1840s James was a farmer for a time near Boddom, a fishing village to the east of Old Deer, but by 1851 he had made his way out to Australia. In that year he married 21-year-old Dianne Wheeler, a recent immigrant to

Victoria from the market town of Wallingford, Berkshire.

Bruce's three Cranbourne parcels bought in 1858 totalled 390 acres. In September 1859 he purchased an additional 466 acres to the west of these Sherwood allotments in Langwarrin, supplemented by 193 acres in June 1860. Over time he acquired a total 3,691 acres into his holding, 'Sherwood-Park,' and built a weatherboard residence of five rooms, along with a dairy, stables, stockyards and piggery.[1] Impressive pine plantations were a feature of the property. The Scot bought McKay's meteorite for one pound, probably sometime in 1860. His intentions were noble; he proposed to donate it to the British Museum, later explaining 'I have spent many a pleasant day in the British Museum, and gained some information.' He wanted to make 'some return.'

Augustus Abel's motivation for his purchase was less altruistic. He was a dealer, familiar with rare metals from his business in gold-rich Ballarat, and saw opportunity in ownership of Lineham's iron. From the duchy of Mecklenburg-Schwerin and a family of renown in creative circles – his father and grandfather were artistic and musical luminaries of the German Confederation – Abel was multi-skilled as a geologist, chemist, and mineralogist. He was also a painter and draughtsman. He journeyed to Victoria in 1858, when well into middle age, and took up residence in Ballarat. His trip with Neumayer was probably a speculative one and, finding McKay's sample already spoken for, he made sure to secure Lineham's.

One can assume Abel travelled to Cranbourne with some expectation of commercial return. His invitation to Neumayer may have been proffered to take advantage of the younger man's expertise in magnetism, useful when examining purportedly iron meteorites. Completing the German complement of the party that day was Karl Rupprecht, a friend of Abel's and proprietor of the Sabloniere Hotel in Queens Street. Rupprecht supplied the cart with which the newly purchased meteorite was transported to Melbourne. His hotel would host Cranbourne No. 2, soon to be alternatively referenced by the name of its new owner, for the next nine months. In early March the *Bendigo Advertiser* would excitedly and incorrectly name 'Mr Ruppricht' as the discoverer of a mass of meteoric iron that 'cannot weigh less than 3000 lb.' Over the following years the 'Abel meteorite' would travel great distances and be displayed in multiple exhibitions while the wrangle over No. 1 (also to be called the 'Bruce meteorite') played out. But Melburnians would be first to get a good look at this prize.

* * *

The Victorian Exhibition of October and November 1861 catalogued a different colony than that celebrated via the Melbourne Exhibition seven years before. Merrett's glass palace of 1854, re-purposed in the interim as the prime concert venue for the Melbourne Philharmonic Society and updated with a water connection from the new Yan Yean reservoir, was again utilised. The now-knighted Redmond Barry once more presided over the list of commissioners, local worthies all. A familiar site and leading names, to be sure, but the colony's population had doubled from even the gold-induced spike seen in the early years of the previous decade, and now stood at over half a million souls. Victoria had changed.

The extensive and enticing catalogue reflected the brash, and justifiably proud, new colony. Its preface remarked on the general disorganisation of the early gold rush years and the paralysis of ordinary industry during those heady times, but also pointed out the now-settled pursuit of the mining sector and the steady attainment of manufacturing and artistic results, fuelled by an energetic and inventive populace. The colony's broad challenges were emphasised in a reminder for the reader:

> *That we inherited nothing of that which we possess; that we have had an immense amount of rough work to perform, in order to render the country habitable, passable, and capable of affording us sustenance; and that we are severed by the circumference of half the globe from the appliances and the civilization of the old world.*[2]

On display were articles representing 'the produce or manufacture of the colony.' The commissioners entreated the people of Victoria, in particular those within the different branches of industry, to 'give them a ready and willing assistance' in supplying a 'collection of objects worthy of the country and of the great occasion.' Selected articles would be transmitted to the International Exhibition of Agricultural and Industrial Products, arranged for London in 1862. Applications were immediate and numerous, requiring an extension to the display floor of 3,200 square feet: work completed in nine days. Exhibits were classified into seven groups, including agricultural products, horticulture, minerals, and machinery.

As might be expected of the world's most auriferous region, gold was on display, and in great abundance. Several Melbourne banks provided selections of their 'crude gold, specimens of extraordinary size, or of peculiar structure, or possessing unusual combinations of rock, ore, or mineral.' The commissioners also intended to send on to the London exposition a crushing machine, to be 'exhibited in action, and quartz taken from different gold fields, at various depths from the surface, will be crushed and washed on stated days.' Victoria's golden halo was to be well burnished.

Abel's meteorite, Cranbourne No. 2, was introduced in the Class IV section, under 'Mineral Products, and the Manufactures and Processes Connected Therewith' as exhibit 121 by Abel, A.T., Ballarat: 'Collection of Minerals and Meteoric Iron.' Even among a gold-rich program this near-pure iron mass had a certain celebrity status, and the Class IV Committee – Royal Society members Frederick McCoy and Alfred Selwyn, and Minister for Mines John Humffray – awarded it a first-class certificate:

> *121. ABEL, A. T., Ballarat. – For a Mass of Meteoric Iron, from Cranbourne, of the greatest interest and of magnificent dimensions. From a scientific point of view this is the most important of our mineral contributions.*[3]

The calibre of the other meteorite exhibitors may have swayed the reviewers, for the governor himself put up a related display. In an echo of the farrier James Scott's 1854 entry, listed as item 129 was 'Barkly, His Excellency Sir H. – Specimen of Meteoric Iron from Western Port, and Horse Shoe made therefrom.' FitzGibbon had given Barkly his samples after the town clerk's Cranbourne excursion. No doubt the governor eschewed a certificate, but another exhibitor won the praises of the judges:

> *153. FOORD, G., Elizabeth-street. – For case of Minerals associated with Gold, and for specimen of Meteoric Iron, beautifully cut in section to show its structure.*[4]

George Foord was a chemist and assayer, later to work at the Royal Mint branch in Melbourne and to join the Royal Society of Victoria. He had come by his sample of meteoric iron from James Bruce, who held a 56-pound block that the blacksmith James Nelson cut from No. 1. Foord divided this piece and etched one of the cut faces, revealing the 'beautiful' Widmanstatten patterns, a striking lattice arrangement often found in irons, to the obvious satisfaction of the Class IV evaluators.

Over two months a steady stream of curious Melburnians attended the articles on display at the exhibition, with weekend days often selling more than one thousand tickets. Other locals favoured the attractions of the racetrack. On 7 November 1861, the first Melbourne Cup was run at Flemington Racecourse, a picturesque location on alluvial flats beside the Maribyrnong River. Local race meetings had been held at the site since 1840, at first in the autumn, with later inter-colonial events showcasing horses from Victoria, Tasmania, Sydney and New Zealand. By 1861 the course had a substantial grandstand and its own railway line and station, bringing spectators right to the track. Race results were telegraphed directly to Sydney. A single-day crowd as large as 40,000 was recorded in 1859, but the first of the later famous Melbourne Cup runnings had only 4,000 in attendance, attributed to the sombre public mood engendered by the news, five days earlier, of the deaths of Burke and Wills.

The risks and excitement of the racetrack reflected the devil-may-care attitude of Port Phillip's boisterous pioneers, an ebullience amplified during the giddy gold rush years. Melburnians could be solemn when the occasion called, but often still displayed the frenzied gaiety of a frontier outpost: a status their city held in only its recent past.

> In the fourth decade of the 19th century the region that would become Victoria entered a new phase of European exploration and, finally, lasting settlement. The Henty family of Portland were among several groups who sailed from Van Diemen's Land in the 1830s in search of opportunity on the continent's south coast. Another concern was the Port Phillip Association, a syndicate of fifteen Van Demonian partners established for the purpose of procuring land from the indigenous inhabitants across

the water. Their party, led by John Batman – a currency lad born in Parramatta – sailed into Port Phillip in May 1835. Batman found the country to be clothed with 'the richest grass and verdure, so delightful to the eyes of the sheep farmer.' Whilst in the vicinity of the Yarra he met with the 'principal chiefs' of the Wurundjeri and gave blankets, shirts, flour, knives and other trinkets in exchange for their marks on what became known as Batman's Treaty.

By August the government in Sydney, recognising no right to title of native peoples occupying land claimed by the Crown, declared such arrangements to be void. In the same month a small party organised by another Van Diemen's Land resident, John Fawkner, unloaded their stores and livestock on the northern bank of the Yarra river and took up residence. By mid-September Batman's and Fawkner's camps were co-located uneasily by the riverside, each eying the other with suspicion. Then, in a move redolent of the hierarchical imbalance of pastoralist and planter in Port Phillip's early years, for £20 Fawkner kept his huts by the landing place but moved his gardens across the river, establishing orchards of apple and cherry and cultivating potatoes, radishes, and turnips in the sandy loam of the south side.

Van Diemen's Land sent more graziers and sheep across the water. In April 1836 the Colonial Office granted approval for a town by the Yarra and for allotments to be drawn up. A sponsored report noted a humble scattering of buildings made from weather-boards, slab, and turf. It was referred to, among other names, as 'Bearbrass' by the locals, perhaps a variation of the Wurundjeri term 'birrarung,' meaning 'river of mists.' By September the Port Phillip District was proclaimed, and Captain William Lonsdale sent down to hold a portmanteau role comprising police magistrate, chief agent of the government and commandant of the district.

Overland incursions began from the north. In 1836 the resourceful Irishman John Gardiner formed a partnership with cattle breeder Joseph Hawdon and Scottish sea master John Hepburn. The trio overlanded 600 head of cattle from Sydney and followed Mitchell's track south of the Murray before reaching the Yarra village in early December. In 1837 South African-born Charles Ebden brought 9,000 head from Tarcutta Creek on the western flank of the Australian Alps over the Murray to his run at Bonegilla and then southward, finally settling on the Campaspe river west of Mount Macedon – the first Port Phillip pastoralist north of the Dividing Range. Englishmen Alexander Mollison made an even longer journey to the burgeoning district, trekking with 5,000 sheep and 600 cattle from his run on the Murrumbidgee all the way to the Coliban river, not far from Ebden's station.

As the pastoral expansion around it continued, the village on the Yarra took on more definite shape. Senior Surveyor Robert Hoddle designed a streetscape to define the settlement and provide a base from which allotments could be drawn up for sale. He proposed a utilitarian but unimaginative grid on the northern side of the river. A cross-hatch of 32 ten-acre plots, with a rectangular military-camp bearing, it took no real advantage from its variations of landscape, such as Burial (Flagstaff) Hill or Eastern Hill. It virtually turned its back on the river and disregarded the south side of the Yarra entirely. Although a huge plot, its one mile by half-mile area set aside no reserves for public spaces. Elizabeth Street was a geometric, geographic, and geologic dividing line. It separated the soon-settled western end, sited on basalt shaped by long-ago volcanic episodes and crowned by Batman's Hill, and the manna gum-dominated bush of the eastern section, formed above deep sedimentary deposits and rising to its own, less pronounced, hill. Mother Nature cast a shallow valley along this boundary, which became a watercourse when fed from rainfall to elevated areas north of the settlement, and it would regularly flood and turn to bog.

In early March 1837 Governor Bourke and his entourage proceeded to Port Phillip in HMS *Rattlesnake*. Lonsdale formed a welcoming party with local notables in attendance. Amidst settlers, gathered Woiwurrung, and a multitude of tree stumps Bourke formally named the raw young community Melbourne, after the serving Whig prime minister. He also took it upon himself to entitle the streets of Hoddle's rectilinear grid. Except for Stephen Street, later changed to Exhibition, his designations have survived to modern times.

The first land sale in Melbourne was conducted on June 1st, 1837, with five of Hoddle's primary blocks being subdivided into 20 half-acre allotments. Batman and Fawkner bought several lots, each securing valuable sites on Flinders Street. Purchasers in these early sales would go on to make impressive stag profits in subsequent auctions, held every six months until the whole grid had been offered up.[5] Auction days were often sodden events; organisers supplied champagne in liberal quantities, and prospective bidders enthusiastically and immoderately consumed it. In a precursor to gold rush excesses, the streets of town and surrounding areas were littered for years afterwards with the signature olive-green bottles.

The appearance of local clans in and around early Melbourne added to the frontier atmosphere. Bare of foot, cloaked in animal skins, uttering a strange tongue, and often smeared with animal fat and red ochre, they were an exotic attraction to the British newcomers. Occasionally, violent encounters between Kulin groups troubled the outpost's perimeter, and despite the hectic activity and developing industry in the young township there was, in those early years, a sense of perilous existence inside a swirl of Aboriginal experience.

Fawkner and Batman developed a squabbling rivalry. During the latter's frequent absences in Van Diemen's Land his younger brother Henry, for a time the settlement's Chief Constable, was a convenient stand-in for Fawkner's antipathy. As well as planting gardens and orchards, Fawkner was a hotelier and newspaper proprietor (when he bothered to obtain a licence). He used his mastheads, first the *Melbourne Advertiser* and later the *Port Phillip Patriot*, as expedient rostra from which to hurl his invective. In his private journals he was even more caustic, noting in March 1836:

> *(Henry) Batman is utterly unfit to hold office here, for he is devoid of intelligence, a mere bore who goes about with his short pipe in his mouth and always drunk when he can get the liquor and is very brutish in his address those times – when sober he is a specious hypocrite.*[6]

The teetotal Fawkner, for years a purveyor of spirits in quantity to parched Vandemonians and eager Port Phillip colonists, knew a thing or two about hypocrisy. In any event, his more youthful opponents were not long for this world – by May 1839 John, disfigured and crippled, succumbed to the syphilis which had blighted his final years, and then Henry was lost to drink, dying in October the same year. Fawkner, forever irascible and argumentative, lived a long and colourful life, taking up land to the north-west where he located his family seat – Pascoeville, the later Pascoe Vale – and a large orchard, and styled himself as the pre-eminent founding father of the settlement.

* * *

The favourable showing of Abel's meteorite at the Victorian Exhibition made for good publicity, and local newspapers also pitched in. The *Herald* of 12

October 1861 reported it 'one of the most interesting objects in our colonial Crystal Palace' and its correspondent described 'sulphurets' running through the mass. Upon cutting and polishing a section to expose the Widmanstatten patterns, a surprisingly patriotic image was revealed:

> *One of the sulphurets, viewing it to the left, shows plainly an image of the Queen, whilst the same on the opposite side presents a likeness of King George III.*[7]

Its putative connections to royalty notwithstanding, the specimen's advertisement was already under arrangement by the canny Abel. A note dated 28 August 1861, written in German and intended for his nephew Professor Frederick Augustus Abel of Woolwich, set out conditions for its proposed sale to a European buyer:

> *Abel's meteoric iron mass of Westernport, in one piece, in size ca. 6 cu. ft. @ 500 lbs, i.e. 3000 lbs in weight, is for sale subject to the following conditions. Buyers for London must leave it at the exhibition here and also, for the duration of the one there at their own expense and risk. The purchase price is £300. The buyer undertakes, upon return of a segment up to 100 lbs, to pay £1 for every lb. This return must, however, take place within a year.*[8]

Abel's price seems arbitrary but expedient. For his 3,000 lbs sample he was requesting a conveniently round number of £300, or two shillings per pound weight. Nevertheless, the meteorite was in demand, and the players that would comprise the cast of the ensuing drama were beginning to assemble. Joseph Milligan was a Scottish surgeon who had spent thirty years in Van Diemen's Land and was now returned to London. The principals of the British Museum engaged him to approach Ferdinand Mueller of the Royal Society of Victoria – to see if No. 2 could be secured for their collection.

The British Museum was the leading such institution in the empire. Established by an act of parliament in 1753 and based upon four foundation collections, notably one bequeathed by Ulster-born doctor Hans Sloane comprising 71,000 books, manuscripts, antiquities, and natural history specimens, it was the world's first national museum. Its trustees acquired Montagu House, a grand mansion of French stylings in Great Russell Street,

Bloomsbury, and first exhibitions were opened to the public in 1759. A stream of donations and bequests expanded collections significantly and required the demolition of the initial building for construction of larger premises in the 1840s. By the late 1850s the museum and accompanying library, now a quadrangular neo-classical edifice, curated an ever-expanding ensemble of antiquities, art and artefacts, ethnographic collections and natural history items. Sponsored excavations abroad became commonplace.

In this period the superintendent of the museum's natural history department, Richard Owen, and his appointee as Keeper of Minerals, Nevil Story Maskelyne, were making concerted efforts to acquire meteoric samples from the various outposts of empire. Maskelyne was a geologist and, later, politician. Oxford educated, he taught chemistry and mineralogy at the University from 1851, taking the chair in the latter discipline in 1856. After his museum appointment in 1857 he catalogued its previously neglected collections by reference to the crystallo-chemical system, and intended to compile the world's finest collection of meteorites. But competition was at hand from museums in Paris and Berlin, and the Vienna Cabinet. British colonial administrators were requested to send collected meteorites to the museum, where Maskelyne and his Oxford students carried out their various research works, including his innovative method of study using thin meteoric sections. Maskelyne instigated a series of exchanges with other collecting institutions, whereby fossil casts and aerolite duplicates were offered for local specimens. The Cranbourne finds would have appealed greatly to him.

In addition to his medical skills Milligan, a long-serving member of the Royal Society of Tasmania, was a geologist, naturalist and explorer who had collected specimens for William Hooker, custodian of the Botanic Gardens at Kew. He seems a logical intermediary for the museum's Mueller request – he was earlier introduced to the botanist by a mutual acquaintance, the naturalist Ronald Gunn, and had been in correspondence with Mueller since 1858.

Ferdinand Mueller was to play one of two pivotal roles in the wrangle over the Cranbourne meteorites. From Rostock, in the Duchy of Mecklenburg-Schwerin, after studying botany at Kiel he emigrated to South Australia in 1847 in the company of his two sisters. Tragically, their parents and five siblings had died of tuberculosis and they were advised to seek a warmer climate. Mueller found work as a chemist in Adelaide but also spent periods on the land, purchasing a small holding at Bugle Ranges on which he built a wattle-and-daub cottage. From here he sallied forth on collecting forays across the Adelaide Hills. He sent papers to Europe on the local flora he sampled, the beginnings of a life-long exchange with peers on the Continent, before Victoria's golden lustre drew him across to the new colony in 1852. There his letter of introduction from Adelaide grazier-politician Francis Dutton brought him to the attention of the now-Lieutenant-governor, Charles La Trobe, who recognised his abilities, referring to him as 'an honest looking German' in a letter to a friend. So impressed was La Trobe that he arranged a sum on the following year's estimates for a new position of Government Botanist. Mueller helped his cause by ensuring La Trobe was made aware of Swiss botanist Carl Meisner's application of the lieutenant-governor's name to a genus of native flowering plants, in recognition of his collecting efforts. Mueller was appointed to his new role in January 1853.

He immediately set out to learn his territory. A willing explorer who had already perambulated widely during his time in Adelaide, the young botanist's first destination in Victoria was the hitherto unstudied alpine region. In a lengthy survey that took in Mt Buller, Mt Buffalo and the upper reaches of the Goulburn and Ovens rivers, Mueller began his familiarisation with Victorian flora and commenced an extensive collection of dried specimens. Returning to Melbourne via Port Albert and Wilsons Promontory, his five-month walk took in 1,500 miles and opened the door on the classification of indigenous vegetation. La Trobe was delighted. After the disappointment of English naturalist William Swainson's posting to the previous office of Botanical Surveyor in 1852 – William Hooker declared him 'as ignorant as a goose' in this field of study – the lieutenant-governor felt completely justified. 'My clever little Botanist has returned having done quite as much as expected & more than any but a German, drunk with the love of his Science, – & careless of ease – & regardless of difficulty in whatever form it might present itself – could have effected in the time & under the circumstances.'[9]

A subsequent and even longer expedition was made to the Grampians, and the Murray and Snowy Rivers, over the summer of 1853/54. Mount Wellington, Mt Bogong, and Mt Kosciusko were tackled via Gippsland twelve months later. By the end of these three noteworthy journeys Mueller was deeply acquainted with the plants of his adopted colony and, by correspondence and supply of duplicate specimens, had begun a lifelong association with Hooker and his son Joseph.

Along with his wide-ranging field work Mueller contributed to the academic and scientific dialog of the colony. He was a founding member of the Philosophical Society of Victoria and a councilman at the Victorian Institute for the Advancement of Science, presenting papers at both learned societies prior to their merger into the Philosophical Institute of Victoria. He was a commissioner for the Melbourne Exhibition of 1854. Given his academic activities and exploratory experience Mueller was appointed botanist to the North Australian Exploring Expedition in 1855, a British government initiative under Augustus Gregory. This 18-month journey began with a voyage from Sydney to the mouth of the Victoria River on the continent's north-western coast. Travelling upriver the party penetrated the Great Sandy Desert before striking eastward and traversing the north of Australia to Moreton Bay; a remarkable 5,000-mile journey during which Mueller observed 2,000 species and identified 800 new items for the Australian botanical lexicon. Additional to his duties as Victoria's Government Botanist, in August 1857 he began a 16-year tenure as director of the Botanic Gardens in Melbourne, and was also appointed director of the Zoologic Gardens.

Mueller was a prodigious letter writer, sending thousands of missives in the course of his professional life. They reflect a conscientious and intelligent man, one well networked with other scientists around the world. Joseph Milligan's approach was a not-unusual request upon his time and labour from a familiar connection within that community. Unpresuming and kindly by nature, Mueller did not refuse a gentlemanly request if it was in his power to accede. He set great store by values of fidelity and loyalty, and as his prolonged solo endeavours in the Victorian wilderness attest, there was a tenacity to his character as well. The British Museum had chosen the ideal channel through which to launch its bid.

4
A Somewhat Eccentric Man

What shall we say, old madman, of this old iron,
Meteorite on a lonely Australian reef
From the age's whirling planet of hope and grief?
Douglas Stewart 'Old Iron'

Some collisions in the Asteroid Belt have spawned such a volume of debris that multiple meteoric specimens on Earth are recognised as belonging to the same parent. Members of these asteroid 'families' have similar orbits and compositions and comprise a majority of the inner belt objects. Their parents are commonly large asteroids, Vesta and the bright Flora among them, which are subjected to a major cratering event, or shattered altogether, by a catastrophic collision. Although more than one hundred families are thought to exist there are only five declared 'prominent,' all identified by the trail-blazing Japanese astronomer Kiyotsugu Hiramaya, active in the early decades of the 20th century.

Asteroid families have an anthropological taxonomy, and include clans, clumps, clusters, and tribes among their classifications. There can be skeletons in the closet; 'ghost' families are remnants of larger, older families long ago broken up and dispersed, ancient and mysterious, and defy standard attempts at identification. A family may also have its share of 'interlopers' – the term describes an asteroid nestled comfortably in the orbital plane of a known family, but whose spectroscopic profile betrays its position as distinct from that grouping. The orphans of the asteroid milieu are those non-family bodies that form a dense and rocky setting for their more illustrious brethren, and suffer the prosaic label of 'background' asteroids.

In time, asteroids and meteoroids may be captured by Earth's gravity.

On vary rare occasions a body may be wrangled into a partial orbit of our planet and linger for a year or two of inconsistent revolutions before slipping its moorings and heading back into space. The Asteroid Belt's rock-and-metal bricolage is the origin of almost all meteoroids that reach the surface of Earth, there to be named meteorites, and they come in great numbers. To be sure, the majority are no bigger than a grain of sand and vaporise on contact with the atmosphere as flashing meteors.

A similar fate befalls the friable composite matter in the tails of those other space wanderers, comets, when Earth's orbit periodically takes it through their trailing wakes. Meteors that catch our attention usually arrive at night, and their high-velocity impacts (somewhere between 11 km/s and 72 km/s) with Earth's atmospheric envelope generate such ram pressure and heat that they leave a visible trail of gases and melted components. Shooting stars have thrilled humans for as long as we have looked to the heavens.

News of the Cranbourne specimens stimulated appreciation of meteorites among the Victorian public, and other noteworthy celestial events ornamented this period, advancing general awareness of astronomy. Between 1857 and 1861 the colony was treated to a solar eclipse, two great comets, a transit of Mercury, and a spectacular solar storm (the last being more of a threat than a treat, but the phenomenon was little understood at the time). In late 1858 a famous object illuminated the skies above Victoria. The Great Comet of that year, also named after Giovanni Donati, the Italian astronomer who first reported it, was the brightest such phenomenon since 1811. It displayed a scimitar-like tail of dust with a straighter gas tail, and numerous artists of the day captured its appearance. Scientist Ludwig Becker produced a lithograph displaying the sweeping apparition over a darkened Flagstaff Hill, with the distant You Yangs lit by a pale October moon and Hobsons Bay a forest of ships' masts. The naturalist Alfred Wallace (from an island in the East Indies), and presidential aspirant Abraham Lincoln (from the porch of his Illinois hotel), were other notables who viewed its passing.

Australian colonies had their share of clever young people. Edward White was a Bristol-born immigrant seduced to Victoria by the siren song of the gold rush. He made meticulous observations of Donati's comet from

Bendigo, where he worked as a mechanic on ore-crushing machines by day and monitored the sky at night. He described his method in a letter to the *Argus* that November, including the 'simple' reflective sextant readings and star-based triangulation process he followed to register the comet's right ascension and declination. This also involved an inverting telescope and the noting of mean-time, or taking of altitudes, so not quite as simple as the autodidact White averred. His letter was published below another on the same topic from the Government Astronomer Robert Ellery, who would subsequently employ him at the Melbourne Observatory. Under Ellery, White and other assistants would make the Melbourne institution, from 1863 merged with Neumayer's magnetic observatory and based in new buildings at the Domain Parklands, one of the foremost observatories of the world. It was the meteorological and astronomical centre of the colony, responsible for weather forecasts, collecting tidal and magnetic data, and maintaining an accurate time reference.

Another self-taught astronomer, the remarkable John Tebbutt of Windsor, New South Wales, made a name for himself in May 1861 with his identification of the Great Comet of that year. Using only a sextant and an ordinary marine telescope, the Australian-born Tebbutt made observations from the veranda of his father's house that were equal in quality to, or indeed better than, those of the official observatories at Sydney, Melbourne, and Adelaide. He described his early May discovery of a faint nebulous object in a letter to the *Sydney Morning Herald*, which was published on 25 May, his 27th birthday. 'I did not ascertain its cometary character till last night, when I found that it had moved about half a degree from the position it occupied on the night of the 13th,' he wrote. Continuing his observations and computations of orbit, in June he made a surprising prediction: the Earth would pass through the comet's tail. The body was closest to the Sun at the end of the month, and a day after this perihelion it had its closest pass to Earth, which was indeed within the tail of the long-period comet for several days, and Europe experienced breathtaking meteor showers.[1]

In early November 1861, the Adelaide-based Observer and Superintendent of Telegraphs, Charles Todd, recorded a transit of the planet Mercury. The energetic Todd, later the architect of the overland telegraph between Port Augusta and Darwin, reported in detail to the *Argus* his observations of the transit on 8 November. He noted it was 'a phenomenon which, though not of so rare occurrence as a transit of Venus, is nevertheless extremely interesting.' Todd had experience of this transit's 'pair,' which occurred in November 1848, in his first year at the Cambridge University Observatory. In typically understated manner he described his accurate observations of tiny Mercury's first and last contact with the blazing disc of the Sun as 'tolerably close approximations.'

To the appearance of comets and meteorites in Victoria's colonial history can be added another cosmic phenomenon – the solar flare. In the last days of August 1859 just such an occurrence, of immense proportions, generated a coronal mass ejection and drenched Earth with a storm surge of electromagnetic radiation. This plasma's interaction with our planet's magnetic field caused a geomagnetic storm of such intensity that vivid auroras, commonly a high-latitude spectacle, were seen as far north as tropical Queensland. Known as the Carrington event after the British astronomer who recorded the visual solar emissions from his Surrey observatory, it was the first time such a display was observed, and extensively documented, from locations in both the northern and southern hemispheres. Georg Neumayer and his team logged the magnetic and electrical turbulence in great detail from the Flagstaff Observatory, noting:

> *The disturbances assumed so violent a character that the intensity at times, and the inclination very frequently, could not be registered, the scales being out of the fields of the telescopes.*

Although the severity of the disturbance was 'such as to render it impossible sometimes to read the instruments exactly,' the Flagstaff crew did just that, Neumayer later publishing a long list of observations. The Aurora Australis was quite a show over several nights, extending into the first week of September; comments such as 'rosy arc from east south east,' 'splendid aurora,' and 'red streamers, very bright,' stand out among the mundane enumeration of declinations and intensities. Observatory staff collected reports from other locations around the colony, including Ballarat and the Brisbane Ranges. In an entry from a William Foy at Shortland Bluff, by the Port Phillip heads, the correspondent found no hyperbole would suffice for the sights he experienced:

> *12:10 a.m. A sudden change took place, surpassing in grandeur any of the previous changes. A dense red centred at the zenith, which suddenly shot out with rapid extension in which streamers intermingled with red to the horizon, spreading from west to east around south, having so magnificent an appearance that I know nothing to compare it to.*

Carrington's stunning electrical distortions also had a significant negative impact on the emerging telegraph networks of the world, including the lines of New South Wales, South Australia, and Victoria. If a solar-terrestrial event of similar magnitude were to occur in the modern era, with our heavily interconnected electronic infrastructure, it would cause catastrophic power outages and damage estimated in the trillions of dollars.

* * *

On 25 September 1861, in a letter to Sir Roderick Murchison of the British Museum, Ferdinand Mueller made note of his findings on No. 2. Murchison had recently sponsored Mueller's application for appointment to the Royal Society of London, and the younger man was careful to show his gratitude. 'When recently I did myself the honour of writing to you, I expressed my warmest thanks for having from you, as a member of the Council of the Royal Society, enjoyed your support for my election.' He then explained the status of Abel's meteorite:

> *It has been secured for local exhibition in Melbourne and after at the London Exhibition, where with many other articles of interest and utility it will be forwarded probably from here in December next on expense of our Exhibition Commission. After the aerolite has been exhibited at your grand show next May, it will be purchasable and I believe Mr Abel will ask through his nephew, Professor Abel of Woolwich, to act as agent for him.*

In fact, the nephew's view of the uncle was not altogether complimentary. Mueller then finished:

> *I thought it but right, that you, my dear Sir Roderick, as ever the head of the British (and the colonies) Geology would be informed of these facts in time, in order that if you thought the piece was in proportion to the value and if the aerolite was desirable for your great national institution, it might be timely secured.*

Murchison held the role of Director-General of the Geological Survey and was in the process of solidifying his eminent status among the scientific elite of Europe. Scottish-born to a wealthy surgeon father, he trained for the army and saw action in the Peninsular War before leaving military service to marry in 1815, at age 23. Rather inclined to the idle pursuits of horse-riding and fox-hunting, at his wife Charlotte's suggestion he developed an interest in geology. He took to the new discipline with eagerness, and his wealth and connections eased his entry to the Geological Society of London and the Royal Society. With Charlotte's assistance he undertook field trips in England, Wales, and Scotland, often in the company of renowned geologist Adam Sedgwick, and spent time on the Continent and in the Urals. He made his reputation with his 1839 work 'The Silurian System,' in which his stratigraphical research provided the definition of this geological period. In time he became a trustee of the British Museum.

Murchison took an interest in the propagation of gold-bearing deposits, and in 1844 made a vague prediction of likely gold finds in the Great Dividing Range of Australia, based on his field work in the auriferous Urals and comparisons with Strzelecki's research. The subsequent rushes of the early 1850s further cemented his reputation amongst geologists and assured his appointment as Director-General of the Geological Survey in 1855. In this capacity he had a role in Australian mineral exploration and offered sage advice on the conduct of that activity. But Murchison was pessimistic of the Victorian goldfields' long-term viability, believing in the primacy of alluvial volumes and that productive seam-based deposits reduced with increasing depth. Australian-based geologists would later dispute his views.

By 20 November 1861, Murchison was in possession of Mueller's letter, and wrote a short note to Nevil Maskelyne:

> *Here is a prosy letter from Dr Mueller the botanist of Victoria*

> *respecting an aerolite which is to be submitted for exhibition here at our great coming show, and then to be sold for £300 (more or less).*[2]

The brevity of the note indicates its likely status as a follow-up to a previous dialog conducted by the two men about the Abel specimen. Equipped with Mueller's information, Maskelyne contacted Augustus's nephew, Frederick Abel, Professor of Chemistry at the Royal Military Academy, with questions, including that of asking price. Frederick, thirty-four years old at the time, had continued his talented family's tradition of achievement. His position at the academy was a prestigious one, and previously occupied by the pioneering giant of electromagnetism, Michael Faraday, for 23 years. In later life the younger Abel would be plied with honours, including a baronetcy, for his work on the chemistry of explosives. But it seems his uncle had displayed some unconventional behaviours. His letter of reply to Maskelyne, dated 9 December 1861, provides an insight into No. 2's owner:

> *My dear sir, I am sorry that I can give you no information on the subject of the meteorite which you say is coming from Melbourne. My uncle, the mineralogist, who is a somewhat eccentric man, went to Melbourne three years ago without letting any of his friends know of his plans – and he has not communicated with me or anybody since his departure. From what I remember of his former business dealings he is not very likely to make a reduction in price, when once he has fixed a price.*[3]

It appears Augustus Abel's instructions of 28 August had not yet reached his nephew. And now the British Museum and its proxy in Australia, Mueller, found other parties to be interested in Abel's exhibit. In Melbourne, Professor Frederick McCoy was making inquiries about obtaining the meteorite. He first wrote to Abel on 16 December, asking to communicate 'on the subject of the price of the meteoric iron from Western Port belonging to you in the Exhibition Building in Melbourne.' In dictation to his secretary he made his case for local retention, downplaying the likelihood of buyers originating overseas, and asked for Abel's best price:

> *As this would be of little or no interest or value in Europe or America where in all museums meteoric irons are abundant, but on the other*

> hand from local associations it would be most appropriately placed in the National Museum of the colony. Professor McCoy is desirous of ascertaining from you the lowest cost at which it could be acquired for the National Museum if you are willing that it should be placed there.[4]

McCoy, like Mueller, loomed large in the early scientific life of Victoria. Born in Dublin early in the 1820s, he appeared likely to follow his physician father into medicine but after some years of study in this field at Cambridge he turned to natural history, geology, and palaeontology. They remained passions for life. With an easy ability for classification and cataloguing, he organised the significant collections of the Geological Society of Dublin, Boundary Survey, and the Royal Dublin Society. Married early to Anna Harrison, McCoy was constantly balancing his study and publishing efforts with the need for employment in support of his young family. Sadly, of the couple's five children three died in infancy.

McCoy's work with the Geological Survey of Ireland, and authorship of two synopses of fossils, brought him to the attention of Professor Adam Sedgwick and led to an invitation in 1846 to arrange the fossil collection of the Woodwardian Museum at Cambridge. The effort would span eight years. Sedgwick was a leading figure in British geology; the son of an Anglican vicar, he was himself an ordained priest. Notable for his field investigations, he had worked in partnership with the more junior Roderick Murchison between 1825 and 1835 on a ground-breaking study of the 'Transition' sequence of rock strata of the lower Palaeozoic.

As a result of this combined effort and its ensuing paper, Murchison's 'Silurian' system and Sedgwick's separate 'Cambrian' period were classified. But Murchison's later publications, including his magnum opus, 'The Silurian System,' gradually annexed previously agreed Cambrian strata, and gave rise to a long-running dispute between the former collaborators over the boundary of the two systems. Employment of the young McCoy provided Sedgwick with a highly skilled palaeontologist who held expertise in fossil-bearing rock layers. It could lead to the proper deployment of his large collection, accumulated over many years, in clarifying the distinction between the two periods. In 1849 McCoy was made chair of Geology and Mineralogy at Queen's College, Belfast, after which he continued his work with Sedgwick on a part-time basis. And sure enough, based on differing faunas he subsequently recognised in relevant beds of early Palaeozoic rock, in 1852 McCoy identified a discontinuity that validated Sedgwick's claims for a separate Cambrian system.[5]

But Murchison's star was in the ascendency, and the astute young Irishman McCoy could not afford to alienate such a well-connected, wealthy, and pugnacious figure. Sedgwick gradually retreated from the debate and McCoy was represented in some quarters as a blindly loyal subordinate. He continued to seek employment. McCoy had some familiarity with Australia from his papers written on coal formations there, and when foundation professorial positions were advertised at the newly established University of Melbourne he made his application. However, without formal tertiary qualifications he was relying on his considerable experience and ability to qualify. The proposed salary, £1,000 per annum, was five times larger than his Belfast stipend, and would have been compelling. Along with assumed references from Sedgwick and other Cambridge academics and Queens College colleagues, support was provided, surprisingly, from Murchison himself. He wrote a glowing testimonial, indicating to an associate that if he had not done so, 'I should not have acted fairly to an able man.'

The wily Murchison may well have had an ulterior motive. McCoy's removal to Australia would deprive Sedgwick of a highly capable and self-assured assistant, and a main source of palaeontological argument in the Cambrian-Silurian controversy. As it transpired, McCoy was awarded the Melbourne position, further adding to Sedgwick's isolation. Murchison's opposition to any change in nomenclature continued, and it was only after his death in 1871 that the debate was effectively settled with the inclusion of the intermediate Ordovician system.[6]

McCoy, raised a Catholic but now converted to Anglicanism – probably under the guidance of his mentor Sedgwick – arrived in Melbourne in late December 1854 with Anna, daughter Emily and son Frederick. It was the month of bushranger Ned Kelly's birth and scant weeks after the simmering tensions of the Ballarat goldfields erupted into armed rebellion. Led by Irish-Australian Peter Lalor, the miners' uprising resulted in bloody suppression by government troops, including the death of fourteen, mostly Irish, miners. A lithograph of McCoy from this period has him in academic gown and high collar, gazing steadily over the right shoulder of the viewer. Wavy hair and long sideburns frame his clean-shaven face. It's a look of quiet determination offset by slightly pursed lips, a mildly hooded brow and high forehead. He looks *young*. Given our lengthy separation from McCoy's era it is easy to overlook his many early achievements, and his relative youth when accomplishing them. All this in a time when the Irish were often at a disadvantage in academic and professional circles, arising from old prejudices and entrenched animosity. All McCoy's applications to British institutions between 1851 and 1853 – once to the Geological Survey and twice to the British Museum – were unsuccessful, and only Belfast had previously offered him a chair.

McCoy was immediately busy in his new role as Professor of Natural History, with wide-ranging subject matter on which to lecture. Class sizes were small in those early days, and the science curriculum took many years to fully establish, but as virtually the only teacher in this stream at the University his was a hectic schedule. The year 1855 saw the publication of the final instalment of McCoy's major work, based upon his many years of analysis and categorisation of Sedgwick's Woodwardian collection; 'Synopsis of the British Palaeozoic Rocks and Fossils in the Geological Museum of the University of Cambridge.' In 1856 he was appointed to the Geological Survey of Victoria.

McCoy maintained his interest in museum work, and when the new governor, Sir Charles Hotham, chose to withdraw funding for the fledgling natural history collection of the colony, then languishing in two rooms above the Assay Office in La Trobe Street, he saw an opportunity. He persuaded the Colonial Secretary's office to permit the displays' transfer to the University, where they would be placed under his supervision and fill dual roles: as material for a teaching museum and exhibits in a public facility.[7] Despite opposition from the Philosophical Institute and elements of the press, who feared such a removal would mean the effective loss of the collection to the public, by mid-1855 the orderly transfer of specimens was underway. But with purpose-built rooms awaiting their new exhibits, McCoy's impatience got the better of him and he arranged for the balance of the collection to be hastily spirited away to the University's Parkville site. It was an early example of McCoy's propensity for action, often pursued without great regard for consequences, and which lent a certain colour to his career in Melbourne. In this example the outcome was a positive one; in 1860 more than 35,000 people visited the National Museum of Victoria.[8]

McCoy's sounding out of Abel over his meteorite was less successful. His exploration of the vendor's floor price, as seen in his letter of 16 December 1861, was made out of necessary thrift. He was usually required to graft and cajole to obtain new specimens, as money for the museum's collection – now housed in the University grounds for five years – was often in short supply. McCoy regularly argued with Chancellor Sir Redmond Barry over funding allocations. As Abel's nephew had warned Maskelyne, his uncle did not reduce his price, and without a willing buyer in Australia No. 2 was readied for shipment to London and the International Exhibition of 1862.

McCoy had to console himself with the certificate awarded to him by the Victorian Exhibition. But while he had presented an important summary paper to the exhibition catalogue, 'On the Ancient and Recent Natural History of Victoria,' it was not this effort that was rewarded. His treatise's palaeontological arguments verified for the first time that the rock sequence in the Southern Hemisphere aligned with that known in the Northern Hemisphere, confirming the geological column as a global phenomenon. Nevertheless, his first-class certificate was actually presented for 'six cases of Australian insects.'[9]

It must have been a wrench for McCoy to see such a prominent meteorite whisked away from Victoria for lack of an institutional buyer. His vision for museums' utility was that they 'become the most ready and effectual means of communicating the knowledge and practical experience of the experienced few to the many.' The establishment on the University of Melbourne grounds of 'a noble suite of rooms, capable of indefinite extension' for this purpose, and his placement there of the public geological collections gathered by Selwyn during his survey, gave a start to what McCoy called 'a connected systematic view of the science as a whole.' His purchase of an additional 2,000 mineral specimens from Europe and America rounded out the set of terrestrial samples, for what he considered 'the greatest public good.'[10] An extra-terrestrial sample such as Cranbourne No. 2 would have crowned the collection. But McCoy was determined in his pursuit of acquisitions, and now turned his attention to the larger Cranbourne specimen.[11] He would not be so readily rebuffed again.

5
This Mysterious Visitant

Meteor trains tubing their cold way through space.
John Kinsella 'Eschatology'

Meteors are nature's fireworks, but offer more than just incendiary attractions. They are a window to Earth's past, and a key to unlock the history of the Solar System's inception, when some dust and gas accreted into planets but other debris did not advance in this manner. Meteorites, i.e. meteors that land on Earth's surface, have the added allure of cosmic gemstones; rich in rare, otherworldly isotopes and some even holding motes of material condensed in other stars entirely. They remind us that we are not passive participants in the universe, and that the universe sometimes pushes in on us. The danger of solar flare radiation is disguised by its colourful dance with Earth's magnetic field, the beautiful polar aurora. The Moon drags tides to and fro in a daily orbital two-step. And our thick atmosphere is a gaseous husk that stands guardian against most alien objects, but not the really big ones.

In Namibia the Hoba meteorite is the heaviest find recorded, 64 tonnes of iron and nickel. Meteor Crater in Arizona is also named after Daniel Barringer, the businessman who bought the site in 1903. Its forceful irruption, 50,000 years ago, left nothing of the original body except tiny droplets of liquefied metal that splashed up and out, then solidified as they fell to Earth. In Mexico the coast of the Yucatan Peninsula bears the stamp of an asteroid collision so powerful that life on our planet momentarily lost its foothold. The Chicxulub event changed the world's weather, contributed to the extinction of the dinosaurs and the decimation of other species, and heralded a new geological era. It is hypothesised that even our own Moon is the product of a colossal planetary strike on its primary. The impact of Mars-sized Theia sundered the proto-Earth, spawned its satellite via accretion of the collision's resulting rubble, and left our planet with its signature tilt.

Earth endures copious assaults by meteors every day, a flux in the order of 40,000 tonnes. But we need only look to the Moon's cratered surface to visualise the even greater pummelling Earth underwent early in its lifespan, nearly four billion years ago. During the Late Heavy Bombardment phase of the appropriately named Hadean eon, a chance alignment of the gas giants Jupiter – the great sculptor of the Solar System – and Saturn changed the shape of their orbits. This created a gravitational slingshot effect on the Asteroid Belt, and Earth and its moon withstood a withering hail of meteoroids and asteroids for millions of years.

In early 1862 the competition for possession of the Cranbourne specimens began in earnest. Frederick McCoy launched his agitations for No. 1 on 3 January, when he first wrote, via his secretary, to James Bruce. He initially flattered the owner. Then, deploying a similar approach to that used with Abel, called upon him to consider home-grown affiliates worthier of claim to his meteorite than overseas interests:

> *Professor McCoy the Director of the National Museum, presents his compliments to Mr Bruce of Cranbourne and learning from his friends Mr Daintree and Mr Selwyn the enlightened views held by Mr Bruce as to the desirability of the large iron meteorite on his (or the adjacent) land being available for public instruction and information for scientific men in a great public museum, Professor McCoy begs to ask the favour of Mr Bruce's cooperation in securing for the National Museum of Melbourne this specimen which would be of much greater interest in this country from local associations than it would be in the museums or collections of any other country.[1]*

McCoy was dropping names. Selwyn, the government geologist who had earlier examined the Cape Paterson area for coal, was Somerset-born and had experience of British coalfields and Palaeozoic formations. In 1852 he was appointed to the Geological Survey of Victoria and soon made director. Disciplined and meticulous, he generated high-quality maps and reports on the colony's goldfields and beyond, all with the assistance of minimal staff. He was not afraid to challenge authority. In 1857 a minor contretemps developed when Governor Barkly sent Roderick Murchison two reports from a government commission on the state of Victoria's mines, written by McCoy but based on Selwyn's research. These appeared to support Murchison's pessimistic view on deep-seam gold sustainability in the colony. Upon reading Murchison's reply to the commission's reports, Selwyn immediately set the record straight, refuting McCoy's position and asserting Victoria's prospective longevity as a gold-producing region. It was a gentlemanly exchange, conducted in the florid prose of nineteenth-century letters, although Murchison retained a haughty attitude throughout. But Selwyn stood firm before the older man's condescension. His colonial voice was measured and methodical and quoted irrefutable data in support of a well-considered argument. Although couched in deferential language, Selwyn's forthright manner shone through:

> *Had Sir Roderick Murchison ever visited, and personally examined, the gold districts of Victoria, he would have found in them sufficient evidence to justify a very great modification, if not an entire change, in his opinions respecting the probable permanence of the yield and the distribution of gold in this colony, although they may be, and probably are, perfectly correct as regards these phenomena in other portions of the globe.*[2]

And he had a final reminder for the grandee of British geology on the value of local knowledge:

> *In any case I have no doubt that Sir Roderick Murchison will be glad to learn what my opinions are on the subject; as, whether correct or not, they are the result of careful and almost constant observation, both surface and underground, extending over a period of rather more than five years, in the largest gold-producing country in the world.*[3]

Murchison, an inveterate self-promoter, jealously guarded his reputation as a geological authority and was slow to acknowledge merit in dissenting voices. Nevertheless, as a result of Selwyn's exactitude he was made to recognise the ongoing feasibility of Victoria's deep-level mines. Sometimes the tail could wag the dog.

Selwyn, with his governmental role and reputation within the mining community, was an important ally for McCoy, who also referenced Richard Daintree in seeking to curry favour with James Bruce. Daintree was another young Briton enticed to Victoria in the gold rush years who, after failing at prospecting, made a mark in scientific and related disciplines. Selwyn employed him at the Geological Survey of Victoria, and he developed an interest in photography which he put to use in his field work, among the first to do so.

McCoy's letter proceeded carefully. Lest Bruce should think his specimen was wholly matchless, McCoy talked down the unique aspect of the sample – even though it was, at that time, the largest iron meteorite in the world. He held dealers in contempt, considering them venal carpetbaggers:

> *Specimens of this kind generally of smaller size are very common in the museums of Europe and America, but the two Cranbourne ones are the only examples as yet made known in this country and the smaller of these two has already fallen into the hands of a person who acting merely in the spirit of a dealer has offered it to the British Museum and refuses to place it in our own National Collection except for such extravagant outlay as would put it out of the question.*[4]

No doubt believing that any asking price of Bruce's would be out of reach for his museum, McCoy did not even broach the subject, instead appealing to Bruce's sense of public service and asking directly for a donation:

> *Under these circumstances Mr Bruce would be doing a great service to Science in Victoria if he would consent to present the meteorite to the Public Museum, the Government (Museum Department) paying all the expense of excavation and transport if Mr Bruce will have the goodness to permit the work to be done.*[5]

But Bruce would not be swayed by these blandishments. He had already refused an approach from FitzGibbon the previous August, stating in a letter to the *Argus* later in the year that he could not release the meteorite to the National Museum of Victoria; 'I regret I cannot accede to your request, as I bought it solely for presentation to the British Museum.' He went on to explain:

> *Shortly after I came up to reside here, some four years ago, I learned from Mr McKay that this mass of iron was lying on his property. I then purchased it from him for a mere nominal value, on the understanding that it was to be sent to the British Museum. I may remark at the outset, that in any subsequent alteration from this contract I had Mr McKay's full consent.*[6]

There is an inconsistency here. McKay offered FitzGibbon No. 1 during his Cranbourne visit, which fell between October 1859 and June 1860. One assumes McKay would not offer something that was not his to give. Yet Bruce places his purchase of the mass 'some four years' prior to his December 1862 *Argus* letter, indicating *he* was the owner at the time of FitzGibbon's excursion. A more likely scenario is that Bruce, made aware of the nature of the specimen by FitzGibbon's visit to his neighbour's property, proposed his purchase to McKay after that event, and it had transpired prior to Georg Neumayer's reporting of same in February 1861.

Regardless, Bruce's transaction with McKay was conditional. An 'understanding' between the two held that the meteorite was to be donated to the British Museum, and should that 'contract' change, it would be with the consent of both parties. An important point, because in his 13 January reply to McCoy, Bruce would now propose something, unbidden, that he had not offered FitzGibbon – that the specimen be divided:

> *In my answer I inform him that 'I had purchased it for the British Museum, but I would allow him one half provided the Melbourne Museum would be at the expense of removal; that the authorities of the British Museum be communicated with, and an offer made to them of one half, provided they would be at the expense of dividing it.' I concluded by requesting an early decision to allow me time, in case of refusal, to write by next mail.*[7]

His request for an early decision is telling, because McCoy's tardy response set in motion an important set of events. Bruce seemed comfortable directing civic and scientific leaders in the labour and expense of handling his meteorite – removal, division – all the result of a fortuitous handshake and a sovereign exchanged with McKay. He was making the most of his opportunity. While a man of means, and a landholder of some standing, his conduct reflected an individualism – wilful and sometimes unconventional – present among many of Victoria's settler population, wealthy and impecunious alike. Bruce followed in the footsteps of some hardy men and women who had made Western Port their home.

> After 1835 white settlement seeped out from coastal footholds and trickled along the river courses and low-lying plains of the Port Phillip District. The land, tentatively examined for so long, was now subject to dozens of private incursions. English and Scottish squatters formed a vanguard. These were often quite young gentlemen, most in their twenties, and the virgin coast and hinterland provided opportunities that were the making of many. The 'Squatting Act' of 1836 was a pivotal development. For men of means, ten pounds per year for grazing rights was their entry to a nascent landed gentry, something unavailable in Van Diemen's Land or New South Wales, where the best land had already been taken up.
>
> Water availability was the key selector in the placement of grazing runs; the Bass River and Yallock Creek on Western Port, Kananook Creek and Mordi Yallock waterway to Port Phillip's south. In the Geelong area three Scots – David Fisher, James Strachan, and George Russell – sought out land around the Barwon and Moorabool rivers to place their sheep. Charles Griffiths and Robert William von Stieglitz ran flocks on the Werribee River and John Wedge, of the Port Phillip Association, was licensed for seven square miles in the same area (a flood in 1852 destroyed his homestead and killed several family members).
>
> John Aitken held a sizeable tract in the Gisborne area, favouring the Saxon breed of sheep but in time producing fine merino wool. George Evans, one of Fawkner's original pioneers, and the brothers William and Samuel Jackson took up runs around Sunbury, where Jacksons Creek cut through bottom land overlooked by tall bluffs. Alfred Langhorne was a nephew of Captain Lonsdale and obtained a grazing licence for 13,000 acres of volcanic plain at Laverton, watered by the shallow Skeleton and Cherry creeks. On the Maribyrnong Joseph Solomon ran 3,000 sheep and his station included a ford high up on the saltwater river, important for west-bound traffic until a punt service was introduced downstream in the late '40s. Across the Yarra John Gardiner, of overlander fame, was now established on the creek bearing his name. By the end of 1838 the

Port Phillip crown lands commissioner reported fifty-seven squatters in the district.

The grassy basaltic plains to Melbourne's north, west, and south-west were most popular as squatting runs. Less so was the land to the south-east, with many marshy areas, including the huge Koo Wee Rup swamp, acting as barriers to transport and communication. Generally, there was less productive terrain to attract graziers. The main creeks of the area – the Cardinia, Dandenong, Eumemmerring, and Toomuc among them – provided water sufficient to needs in most seasons, but it was not an easy existence. Nevertheless, sturdy types did make their way there.

Among the first were Terence O'Connor, who brought stock across Dandenong Creek in 1836 and eventually held a run on Cardinia Creek, and the Ruffy brothers; five siblings in a family of nine, sons of an English agricultural journalist, who had previously tried their hand in Van Diemen's Land. In 1836 the brothers acquired a run named Tomaque on the plains north of Western Port. Two miles to the east they also held in collective the large Mayune run from 1840 until one of their number, Frederick, assumed the lease of its 32,000 acres outright in 1845. From the Ruffys we first hear of the name later attributed to the township and district; along with their pastoral pursuits they operated the Cranbourne Inn, an ale house possibly named after the Berkshire town.

The issue of grazing licences introduced squatter-led settlement north of Western Port, whose new arrivals were more diversified in their land usage than other settlers in Port Phillip. The wetter climate meant cattle found favour over sheep, elsewhere the dominant stock. And as the 1840s advanced pastoralists around Cranbourne doubled as gentlemen planters, with crops of barley, wheat, and potatoes. Most stations, according to the Nottinghamshire druggist and colonial tourist Richard Howitt, were 'situated on the rich deep soil, almost interminable thick-grassed meadows of Western Port, a fine cattle country.'[8] Still, bushfires, floods, and the occasional escaped convict served to heighten the senses. Wild dogs could be a problem and foot rot had to be monitored.

Stations were often remote, and rough huts of wattle and daub, or turf with a roofing of stringybark, were home for many. Another gentleman author, the Scotsman Robert Murray, sojourned over a summer at Port Phillip in the early 1840s. He described the 'principal hut' of a station he visited:

> *Their whole appearance is characteristic of a half-savage state of existence The walls are constructed of that material known in the colony as 'wattle and dab,' or, in other words, a frame of wicker-work overspread with mud, and support a roof covered with rolls of bark which the wooden stretchers that press them down can scarcely keep from resuming their original circular shape.*[9]

His opinion did not improve upon being invited indoors:

> *We perceive we are in an apartment that seems fresh from the hands of Robinson Crusoe. The inside of the walls differ in nothing from the exterior except that the mud is a little smoother.*[10]

But some enterprising squatters enjoyed comforts familiar to the settled areas. At Tooradin, south of Cranbourne, the Mantons had a fine house with verandas and French windows, and F. M. Mundy enjoyed sheets on his bed on the eastern shore of Western Port. Even Murray's unnamed host owned a sofa and table of mahogany, and indeed a pianoforte, albeit with a missing leg. These mock lords of the land were in fact young custodians of quite large holdings; the Mallum Mallum run, on the smaller side, was 1,600 acres, while Towbeet encompassed 12,800 acres, leaving its custodian Ruffys plenty to administer.

Runs changed hands frequently. The Barker Brothers of Cape Schanck took up Heifer Station, south of Mayune, in 1840. By 1843 it was in the hands of a William Smith, who re-named it Carnmallum (Hugh McKay was overseer on the property, and in 1855 took out a pre-emptive right claim for his 650-acre 'Murrangang' parcel of this run). Towbeet passed from the Ruffys to Sam Webster, then on to Hugh Glass, in the space of four years. Victoria's Lands Commissioner held magisterial authority in this unsettled area, and, amid sundry other responsibilities, issued grazing leases and depasturing licences.

Occasionally tribal quarrels impinged upon daily life, such as when Gunai and Boonwurrung clashed in 1841, terrifying the women of a station on the Bass. The slower spread of whites into Gippsland meant Aboriginal ways lingered longer there. Polish explorer and geologist Pawel Strzelecki changed that with his notable 1840 foray from Yass, via the Snowy Mountains south to Western Port. Although nearly ending in the death by starvation of his party, it marked the beginning of the eastern push by pioneers and the beginning of the end for the indigenous way of life. Ironically, Gundangara man Charlie Tarra, who hunted game for their food when supplies ran low, saved Strzelecki's group with his timely guidance, and brought the party through to Corinella and safety.

For a time, Strzelecki had followed a trail blazed by Angus McMillan, another Scot to have an impact on Victorian exploration. McMillan's 1839 journeys south-west from Corrowong won him plaudits as the discoverer of Gippsland and led to pastoral interest in the lands around its central plains and lakes. But his leadership of subsequent massacres of the Gunai, most infamously that at Warrigal Creek in 1843, none of which he answered for in his lifetime, has blackened his reputation beyond recovery.

***

In October 1861, in addition to Joseph Milligan's earlier inquiry regarding the Abel specimen, Nevil Maskelyne requested Ferdinand Mueller to advise on the mechanics of removing the Bruce meteorite. Mueller wrote to the Mining Department for guidance. But by the time of his late-January reply to Maskelyne he had not received the requested estimates, due to 'difficulties being experienced, as it has to be brought across an unbridged watercourse and I think to be extricated out of a swampy place.' In that letter Mueller showed his colours:

> *I would advise that the influence of the Secretary of State for the Colonies should be solicited to induce the Government of this colony to forward this large meteor to the noble British Museum, where the value of the specimen would be higher appreciated than elsewhere. Some of our population are however very anxious to retain the specimen in the colony, but I believe any application from the authorities of the British Museum supported by the Home Government will place you in possession of the meteor.*[11]

Mueller may have felt that with Abel's meteorite in the wind, the best chance for home placement of a Cranbourne specimen lay with No. 1. He did not hesitate to promote the British Museum as the appropriate location for the meteorite, believing its value would be more recognised there 'than elsewhere.' In his correspondence with Maskelyne and Murchison he regularly inserted positive references to the British Museum, e.g. 'noble,' and 'great.' It seems a little obsequious. But Mueller was well versed in the interchange of information and specimens with overseas institutions, and had a deep network of contacts within the worldwide academic setting. He was recently promoted to the Royal Society of London. Although Melbourne-based, he was a global scientific citizen, and perhaps did not hold the same sense of connection to local artefacts felt by McCoy and, as shall be seen, other leaders of science in the colony.

Meanwhile, Governor Barkly was busy coordinating personnel. Being aware, via FitzGibbon, of Bruce's initial preference for his sample's disposal, he instructed Mueller to take steps for removal of No. 1, and proposed funds for transport expenses. Mueller wanted to visit Cranbourne in person but pressing engagements caused him to send his assistant, Ernst Heyne, down to see Bruce on 3 February. Heyne, one of Mueller's team at the Botanic Gardens and also a mathematician and linguist, carried an offer 'on behalf of His Excellency and the British Museum assistance for the conveyance of his meteoric mass to London.' At this stage Mueller was unaware of Bruce's offer to McCoy to divide the meteorite.

Now McCoy finally stirred. He responded to Bruce on 4 February 1862, thanking him 'for his promise of even half of his great Cranbourne meteorite for the National Museum of Victoria.' He explained that Richard Daintree had been tasked to call on Bruce and clarify McCoy's claim, but this had not yet occurred. McCoy was nervous of his standing with the farmer, as he took pains to remind him that Abel's specimen was for sale to the British Museum, (which was not confirmed at this point), such that that institution would have a 'superfluity' of meteoric samples. He also sought assurance that Bruce would stand by his offer of half his own sample for the Victorian museum:

> *Professor McCoy need only remain satisfied with the promise of the half made by Mr Bruce and when the Estimates are sufficiently forward will complete the arrangements he had already communicated for getting tender for the removal at the cost of the Museum.*[12]

But he was too late. Bruce replied to McCoy on 9 February, explaining that he could now not honour his previous proposal:

> *My answer to your previous letter, in which I offered half the meteorite to the National Museum on certain conditions was posted on the 13th ultimo and as no answer was received up to the 3rd instant accepting these conditions I naturally concluded they were not acceptable.*[13]

Bruce explained that, based on his conclusion, he had instead accepted Mueller's offer, conveyed by Heyne on 3 February, to send the entire specimen to the British Museum, with Barkly's financial assistance. Aware of McCoy's and Mueller's association with the Royal Society of Victoria, he had assumed the two agreed 'as to the steps each was taking in the matter.' In this assumption he was mistaken. But he still held on to the thought of dividing No. 1 between the competing museums:

> *I regret much, that one half of it cannot be retained for the National Museum as I think the merely dividing it in two would not make it less interesting, besides showing its structure without further trouble. I would suggest that you appeal to Dr Mueller and if he will allow it to be divided I shall offer no objection.*[14]

McCoy immediately wrote a lengthy response, on 13 February. He once more referenced Daintree's earlier delay in meeting the farmer to 'personally explain to Mr Bruce the claims of the Melbourne National Museum on his considerations.' He offered this as the reason for his own slow response, and 'not any unwillingness to accept the half if the whole would not be given.' In fact McCoy had not been idle, working his contacts and eliciting support for his cause from Georg Neumayer. In a letter dated 5 February, the German wrote; 'I am certainly of the opinion that the larger Cranbourne Meteorite should remain the property of our Colony, through the national museum, and it would afford me great pleasure to assist you in assaying out this object, as far as it is in my power.'

There is a touch of the tinker about McCoy. He could be mischievous, bordering on roguish. He was being flexible with the truth when he again tried to wedge Mueller in the letter of the 13th, intimating to Bruce that the botanist was the agent of Augustus Abel and therefor responsible for sending that specimen to the British Museum. He reminded Bruce once more of the

Scotsman's offer to divide the meteorite, ignoring that offer's subsequent withdrawal, and held out the imminent prospect of money forthcoming to facilitate its removal and division. And despite his repeated reference to Daintree's previous belatedness, he nonetheless quoted the young geologist's promise of assistance, along with that of Selwyn:

> *On the part of the National Museum therefore Professor McCoy as Director (and armed by the kindly expressed wish of Mr Bruce in his note of the 9th to see half of his specimen retained for the Melbourne Museum) will forward as soon as any money for the year can be got from the Treasury (in a few days) to carry out the arrangement for the transport and division of the specimen Messrs Selwyn and Daintree having kindly promised to aid in superintending the operation.[15]*

Next, another dig at Mueller, this time for his modus operandi:

> *Professor McCoy quite agrees with Mr Bruce that it was natural to suppose that Dr Mueller had communicated with Professor McCoy before proceeding in a matter so much out of his department as to send his Botanist Assistant to secure the specimen but the matter was kept secret.[16]*

Then he invoked the governor and, finally, trained his sights on Owen and Maskelyne:

> *Professor McCoy also states that his Excellency the Governor was also quite unaware of the desire of the National Museum to possess one of the two meteorites when he offered Dr Mueller to pay the cost of transit lest it should fall into Mr Abel or similar hands and His Excellency has previously made some move to secure one at least of these two meteorites for the government of this colony but on Dr Mueller's application gave up his intention.*
>
> *He now writes to say that under the circumstances he approves of the division of the specimen accompanied by casts as formerly mentioned if there be no relenting on the part of those who contemplate the more than ostrich-like glutting of the British Museum swallowing up our Victorian meteoric iron masses when one alone would form fit food for all their philosophic speculations and leave us something to feed on as well.[17]*

Despite his apparent annoyance McCoy took a poetic turn of phrase, invoking a biblical reference via the Book of Lamentations in comparing the British Museum to the perceived cruelty and greed of ostriches. He held a healthy scepticism for the metropolitan agenda, and wanted as much of the Bruce meteorite as could be gotten. Royal Society members Neumayer and Selwyn were poised in support. But on the same day as McCoy wrote his long missive to No. 1's owner, Bruce was affirming his arrangement with Mueller. The Scot was still amenable to the question of division but did not appear to trust McCoy to make that arrangement, preferring the 'greater facilities at home' for this purpose:

> ... *You will see by the letter I sent to Mr Heyne yesterday, that I have placed the meteorite wholly at your disposal, of course there is a wish expressed, that one half may be retained for National Museum, but I certainly think that Prof. McCoy has forfeited his claim on it. I think I was quite justified in thinking my terms were not acceptable and even now he gives me no guarantee that it will be divided, however I leave the matter entirely in your hands.*
>
> *Would it not be as well, (even if you agree to let Prof. McCoy have one half) to send it home to the Exhibition and get it afterwards cut in two, as there are greater facilities at home than here for getting such a job done.*[18]

The next day Mueller picked up his pen and explained to McCoy his desire to keep the specimen whole, even though he had hesitantly agreed to its division. It is easy to feel sympathy for Mueller's predicament, caught as he was between personal friendship and professional propriety. He explained his reservations: 'I was not inclined to act in direct opposition to your wishes, and agreed reluctantly to the division of the specimen,' and 'we are doing a most deplorable act in cutting such a rare cabinet piece.' Ever the diplomat, Mueller acknowledged McCoy's motives but felt they were dealing with an exceptional case:

> *I respect your desire of doing the utmost for the institution over which you so zealously preside, but I cannot but feel, that the scientific men at home, and especially the worthy administrators of the British Museum will censure my action. If I was in your position,*

> *I must confess, that I would take a cosmopolitan view of the matter and wave my local interest (strong as I entertain it towards my own establishment) in an exceptional instance like this, reflecting how much more would be served to science, if such a treasure was rendered accessible to men of science of all parts of the world, in the noblest institution which ever will exist for the preservation of the products of nature.*[19]

Mueller may have felt 'the scientific men at home' looking over his shoulder, for they seemed to be guiding his hand. His motivations are complex. What exactly did he feel when he mentioned 'home' in this context? Perhaps it was a general sense of Europe, rather than the British Empire, given his upbringing on the Continent and his diverse academic networks. It was only in the previous year that Mueller had finally acquiesced, upon Barkly's intercession, to assist the English botanist George Bentham in preparing a conclusive volume of Australian botanical life. This was one in a series of colonial floras sponsored by the Royal Gardens at Kew, a project for which he had long cherished the primary role.[20] With that aspiration extinguished, and reduced to the role of assistant in the great work rather than author, what drove him to act as the agent for another British institution in this matter of the Cranbourne masses? One can surmise that Mueller had the power to contain, even suppress, individual ambition if the greater good of scientific enlightenment was served.

Mueller's enthusiastic promotion of his adopted home and its natural history attractions leavened his cosmopolitan outlook regarding the meteorite. He was always a fervent proponent of Victorian forestry, agriculture, and botany, and later described to Governor Darling his 'patriotic love' for the colony. His advocacy of eucalypts led to their introduction offshore for aesthetic and medicinal purposes, and the development of eucalyptus oil on commercial grounds. As a leading member of the Philosophical Institute and Royal Society he made important contributions to Victorian scientific awareness, and acted as a virtual arm of government inquiry at a time when learned societies were often used to gather data on the colony's resources.

But Mueller's professional life had its difficulties, and the disappointment of *Flora Australiensis* was not the only slight he had to bear. Since his time in Adelaide he had seen his papers published on the Continent with the assistance of contacts there, notably Hamburg botanist Wilhelm Sonder. In the mid-1850s Mueller had also supplied descriptions of new and unusual Australian specimens to William Hooker for proposed publication in the English botanist's scientific journal. This was a reasonable expectation for a colonial government botanist, and the volume of samples dispatched was markedly increased after Mueller's return from the Gregory expedition to northern Australia. But Joseph Hooker eventually returned his manuscripts, citing a lack of time to validate them against known species, a necessary pre-publication step that the German could not perform himself in the days before his herbarium was properly established.

Fortunately for Mueller, the transactions of the Royal Society of Victoria and its predecessor organisations provided an alternative forum in which to publish important papers, and in subsequent years his own scientific journal, *Fragment phytographiae australiae*, became his primary vehicle for publication.[21] If the Hookers' attitude rankled, it did not affect his correspondence with the pair, which he maintained for many years. He looked upon William as something of a father-figure (Mueller lost his own father when a young boy), and his death in 1865 had a particular affect, Mueller writing to Joseph that the news had thrown him into 'an abyss of sadness.'[22] The Cranbourne meteorite issue would mean he later faced another distressing episode in his professional relationships, one much closer to home – with Frederick McCoy.

6
Divisions Over Division

Their music dislodges a meteor from erratic flight,
its burning up against the dark cross of the mountain's folds.
John Kinsella 'Canto of my Great Great Grandfather
– Edward Pat Kinsella'

Meteoroids that land as meteorites present a glimpse of times long past. While our Earth has been warped and weathered, these chunks of ancient planetesimals have remained pristine time capsules, careening through the deep freeze of space. For such bodies to reach our planet's surface at all involves a hefty element of chance. A meteoroid must move into an Earth-crossing orbit. It must intersect that track at the same time as Earth passes. It must strike the atmosphere at the right angle, so as not to 'bounce' off. It must survive entry and make landfall, not splash into the sea. It must last through years of weathering and the biological and geological processes that work to conceal it. Finally, it must be recognised as an interloper among a terrestrial rock background.

Meteorite classification recognises three main categories. The stony class comprises mainly silicates, similar to basalt; irons are mostly iron with some nickel; and stony-irons contain a mixture of roughly equal quantities. Within the stony classification are the chondrites, comprised of chondrules that formed from the melted dust-brew of the early solar nebula, and then cooled to often beautiful glassy beads within accreting bodies. Stony varieties make up the majority of meteorites seen to fall, with irons and stony-irons comprising only 5% of falls. But irons' distinctive appearance means they are preferentially collected, and of all the meteorites found in Australia, irons account for approximately 29% of the total.[1]

> Meteorites are like astronomic emissaries; they can bear messages from primordial detonations. The irons arise from the cores of planetesimals, bodies formed early in the Solar System's life by accumulation of dust grains and other diffuse matter in the solar nebula. After gradually cooling, iron-rich sediment decants under gravity to the centre of the solidifying mass. This body is then broken up in collision with another planetesimal and ejects smaller meteoroids, including heavy, core-sourced varieties rich in iron-nickel alloys.
>
> In 1818, during British explorer John Ross's first Arctic expedition, his party encountered Inuit in the Melville Bay area of north-western Greenland bearing bone-handled knives and harpoons with cutting edges made of iron. The indigenous people described an 'Iron Mountain' as the source of the metal. American naval officer and explorer Robert Peary later found this to be three sizeable meteorites of a probable single fall, later named for nearby Cape York, with the largest weighing 31 tons. The inventive Inuit travelled to this far northern location for many generations, bringing hammer stones of hard basalt with which to wrest workable pieces from the three masses. They would cold-smelt these iron samples to craft their tools and hunting implements. Under the nearby Hiawatha Glacier is a 31-kilometre-wide submerged crater, theorised as the impact site of the Cape York meteorites' parent. In 1894 Peary made off with the smaller two of the three specimens, the 'Woman' and the 'Dog,' returning in 1897 for the largest – the 'Tent.' All were sold to the American Museum of Natural History, where they are on display to this day. The Inuit's thoughts on the appropriation were not recorded.

When Frederick McCoy answered the letter of his friend and colleague, Ferdinand Mueller, on 17 February 1862, it was at the end of a communication dealing with zoological matters and he used a humorous throw-away line to close it out, suggesting:

> *As for our rival claims on the meteorite I would suggest the good old test of single combat and walking on red-hot plough-shares to see with whom the right lies – unless indeed the British Museum would solve it by sending us Abel's one.*[2]

McCoy's proposed test for dominance was intended in jest, of course, but the idea that No. 2 might be exchanged for No. 1 may be viewed in a different light. Despite McCoy's mistaken assumption that Abel's sample was already within the gift of the British Museum, it was a proposal soon to gather momentum. Mueller, seeing his chance, supplied his own impetus. He wrote to Maskelyne on 20 February, explaining events up to that date; Bruce's original intentions, the dispatch of the Abel specimen to London, Bruce's subsequent willingness to divide his specimen and McCoy's acceptance of half. Describing his opposition to division, Mueller went on to represent McCoy's flippant suggestion as something more than a glib aside:

> *Writing to Professor McCoy as you will perceive by the enclosed copy, I pointed out the indesirability of such a course, when my friend agreed that if the Abelian Meteor could be secured by you and returned to the Melbourne Museum, he would forego his partial claims on the larger specimen.*[3]

Using the common practise of inserting copies of referenced letters along with the key message, Mueller ensured Maskelyne was aware of his 14 February reluctance to agree to the suggested division of the Bruce specimen. He may have been embellishing when he wrote that McCoy had 'agreed' to forego his partial claim in the event the Abel sample was returned. However, that confirmation came on 21 February when McCoy wrote to Bruce to explain that Selwyn had arranged with Mueller to remove No. 1 from Cranbourne and 'bring it down' to Melbourne. In this letter McCoy refers to Governor Barkly as being the one who proposed the swap:

> *His Excellency the Governor has made a suggestion to Dr Mueller to the effect that if the British Museum purchases Abel's meteorite and sends it back to the Melbourne Museum there would be no necessity to cut the larger one and this meeting the views of Dr Mueller would probably be the most satisfactory solution of the difficulty for all the parties and as Mr Bruce has put the management of the affair into Dr Mueller's hands his wishes will be followed in the matter.*[4]

On the same day McCoy put in a mild protest to Owen about the British Museum's desire to obtain both meteorites, although without reference to ostriches. But he was still trying it on, building up the size of Abel's specimen ('one of the large samples') and downplaying the size of Bruce's ('the somewhat larger one'). He again mentioned Barkly's suggestion to exchange No. 2 for No. 1, instead of separating the latter, declaring this 'would probably be the most satisfactory solution of the difficulty for all parties.' It would be an important concession. With these two letters he indicated that one complete meteorite was preferable to a meteorite divided or, too awful to contemplate, no meteorite at all.

McCoy enjoyed a scrap, but he also held a passion for the dissemination of knowledge. It would be gathered by what he called 'the experienced few,' so that it could find practical application by 'the many … which depend more or less directly upon a knowledge of the peculiarities of the raw materials which Nature furnishes to us.' Importantly for a museum director in mineral-rich Victoria, he thought of himself as 'one who likes to see science applied to the useful purposes of life.' He arranged for working scale models of Bendigo mining machinery to be displayed. These afforded '"eye knowledge" to a class of persons who have neither the time nor opportunity for lengthened study of books,' in particular 'persons landing here on their way to the mining districts.'[5] McCoy considered this public role of the museum to be of paramount importance, and over the second half of the nineteenth century the National Museum of Victoria was recognised internationally as one of the world's great natural history museums.[6]

Although sometimes pugnacious, McCoy could separate his professional allegiances in a way the ingratiating Mueller could not. Even as he prevaricated over the Cranbourne meteoric specimens, to the potential detriment of the British Museum, he was being generously supported in time and effort by that institution's keeper of zoology, John Gray. Gray was a primary conduit for McCoy to European dealers of natural history items and had already secured collections of high quality for the Victorian museum.[7] But Bruce's sample was exceptional, and the Irishman managed to compartmentalise his competing priorities while its availability was still in play. McCoy could afford to play a long game, for he was about to get his hands on the prized meteorite.

* * *

While McCoy and Mueller jousted via correspondence, in the hot and dry third week of February 1862 Alfred Selwyn gathered a small party in preparation for the extraction of No. 1 from its Cranbourne site. Richard Daintree, the Assistant Geological Surveyor, was involved, as was George Foord, the chemist who exhibited a cut sample, taken from No. 1, at the recent Victorian Exhibition. Enoch Chambers, a founder of Prahran, had won the contract for the heavy work of the meteorite's removal, worth £100. He brought a wagon and tools, and the manager of his engineering works, Benjamin Barnes, to assist.[8] At Selwyn's invitation Georg Neumayer was also among the group that day, joining them at the Brighton train terminus, from where they proceeded to Cranbourne via Dandenong. They arrived at Bruce's on Thursday 20th at midday and Neumayer set about taking magnetic and astronomical readings in the vicinity. By the evening their equipment was established at McKay's neighbouring property and a camp set up close by the meteorite.

In the morning Neumayer made a series of measurements with his magnetic theodolite from stations gradually approaching the specimen. He used a chalk to indicate the line of intersection separating the iron body's north and south magnetic 'fluid.' Foord noted 'the portion above the surface powerfully attracted the magnetic needle, and had local poles, so that in passing the needle over it the latter swung around, and became reversed in particular parts of the meteorite.'

No doubt Chambers and Barnes were eager to get started, but even with the young scientist's careful activity taking a portion of the early morning they made rapid progress. Neumayer reported:

> *After the earth around the meteorite had been so far removed as to allow of the machines required for the raising of such an immense mass, to be applied, it did not take long to show that the undertaking of this arduous task would prove a complete success. It was about 10h in the morning of the 21st of Feb. when, by the aid of a screw-jack, the mass made its first motion since its arrival on our planet.*[9]

As was his practice, Daintree took photographs of proceedings during the removal, two of which survive. One is a view down to the *in situ* meteorite, taken a few steps from the yard-deep access hole in which it sits exposed. Pick marks score the sides of the pit. The background is shaded by a nearby tree and fallen timber clutters that space. One broad side of the fragment faces up toward the camera and flares a bright pale within the otherwise grey shades of the composition. A protuberance, possibly the section truncated years earlier and the basis for the horseshoe, can be seen at the rear. A linked chain is draped over the top face and runs up one sloping side of the excavation, across some logs and out of view in foreground.

The other picture is from a similar aspect but closer and lower down, almost at ground level, from later in the day. The brightness of the exposure is more uniform than the first, so the meteorite looks to be made of the same earthy materials as its surrounds. Neumayer's chalk marks are visible. Now the chain is trussed tightly around the top and sides of the specimen, which has been inverted through 90 degrees and appears to teeter, top-heavy, on its broad point. To its right it is braced by a leaning screw jack, used to raise it into position. One is reminded of a captive bear; hefty, chained at paw and snout, readied for severe entertainments.

As seemed customary for those in contact with the Bruce meteorite, Selwyn filched two small samples from its base before it was carried to Melbourne in Chambers' wagon, a three-day journey. There, in the grounds of the University, it was positioned in front of the National Museum of Victoria, whose mineralogical collection already exceeded its available display space, and was aligned on the same bearing it held when in the ground. Visitors were accepted daily, Sundays excepted. If possession is nine-tenths of the law, then Museum Director McCoy was well-positioned for the ongoing custody battle. The *Herald* reported in November:

> *The Museum, however, is arrived at by the straight path leading to the University from the entrance gate. Near to the circular flowerbed at the top of this path, within a wire enclosure, is the famous Bruce Meteorite, which fell at Cranbourne, near the Dandenong road.*[10]

When passing through the Cranbourne area again, in January the following year, Neumayer could not resist making a flying visit to McKay's, to see once more the place where the meteorite had lain for so long. He noted 'The place now looks very deserted, a simple water-hole alone marking the memorable spot.'

In a letter James Bruce wrote to his brother in March 1862 he indicated some difficulties were experienced with the transportation of his sample. Hardly surprising, given a bullock dray was the means of movement for a cargo 3.5 tons in weight, of irregular shape, and compacted into a volume of approximately 17 cubic feet. Bruce airs his thoughts on the question of dividing the meteorite:

> *The Meteor at last got to Melbourne after some stickings on the road including smashing a timber waggon – it will not be at the Exhibition although its smaller sister may – it is not decided yet whether it is to be divided or sent home whole it will depend on the decision of the Authorities & the British Museum there has been some correspondence about it amongst our scientific men & the whole including one letter from me I think will be sent home to show the opinion here as to whether it should be divided almost everyone including some turncoats are for a division as it would show the inside structure this has been my opinion all along & I had to fight it against some that were then of a different opinion but who have turned round now & the only reason that it is not divided at once was because I made a promise of the whole to the British Museum & unless they or their representative give it up of course it will go home whole as a matter of course the British Museum is a much fitter place for it but I think if divided both parties would be saved & the meteor would at the same time be more interesting.*[11]

Bruce's lack of punctuation makes his grammar a challenge, but he is plainly dissembling on the matter of the specimen's division. Showing 'this inside structure' had apparently been 'his opinion all along,' and he even fought for partition with an element who favoured keeping the fragment whole (these 'turncoats' later favouring division). However, immediately thereafter in his concatenated sentence he explains his earlier promise of the *whole* meteorite to the British Museum. The contradiction is clear: he could not have favoured division 'all along' if he had also proposed the shipment of the undivided sample at one point.

* * *

While Bruce maintained this misleading crotchet, the question of division continued to percolate. Selwyn, having overseen the removal of the meteorite from its resting place, sought an opinion from Foord on the scientific advantages of halving the specimen, 'so as to display the internal structure and composition.' He also asked him to consider whether forwarding it intact to the British Museum, or sending a model plus one of the halves, was a better option. The assayer was firm in his response of 27 February, believing 'the only possible mode of making out the chemico-mineralogical nature of this meteorite is by cutting so as to display the interior.' And he was not referring to a simple section of material, such as he had taken from Bruce's block and exhibited the previous year, but that:

> *It is desirable that the section effected should bring to light as long a surface of the interior as possible, the mass being comparable to a new and unexplored country in any part of which a discovery may be made. The surfaces thus displayed and etched would prove not only of the highest scientific interest but, in any museum in which they were placed, eminently more instructive than the intact encrusted mass.*
>
> *… Gaining these advantages by cutting the meteorite I cannot see that any useful purpose would be lost. A faithful model of the entire mass with particulars of its perspective would fulfil all the purposes which the intact mass itself could observe.*[12]

Selwyn made a similar request of Neumayer, who provided thoughts on the matter in his own letter of 27 February. Always a diligent analyst, the young German recommended thoroughly weighing, measuring, drawing, and photographing the specimen. Like Foord, he believed sectioning the meteorite to be an 'exceedingly valuable' task, going even further when he suggested using pieces of Cranbourne No. 1 to assist exchange with other museums in order for Victoria to be 'assured an opportunity to acquire a complete collection of meteor fragments.'

The forces for division of No. 1, and therefore its part-retention, were gathering around Selwyn. But Mueller continued to hold out. He wrote to Selwyn, also on 27 February, and formally protested division 'on behalf of the

British Museum against any such measure.' He reiterated his understanding of McCoy's position regarding the return of the Abel specimen and explained that he was awaiting Maskelyne's and Owen's approval of the swap. Mueller had earlier stated a preference for London as No. 1's proper home, being where it would be 'higher appreciated than elsewhere' by 'men of science of all parts of the world.' He also believed that 'such scientific experiments, as will likely be instituted in London on this large meteoric metal, cannot be of equal value after the division of the meteor.' Furthermore, and with the premier status of the meteorite in mind, he wrote, '… since divided it would no longer surpass all others of the series located in the British Museum.'

Perhaps the assertion that experiments on a divided 3.5-ton specimen were less worthy than any conducted on its complete mass is an example of the botanist straying too far from his area of expertise. The quick-tempered Selwyn was sharper in his tone and language than Mueller. In his 28 February reply the geologist wrote, 'I have no objection whatever that the course you propose should be adopted in reference to Bruce's large meteorite,' continuing:

> *but I must repeat that I differ entirely with you as regards the propriety in a purely scientific point of view of having it divided, and I wish you to understand, that whatever its ultimate destination, either whole or in part might be would not alter my opinion in the matter, because I consider that by cutting it in half very important results may be expected as regards our knowledge of the physical and chemical properties, particularly the internal structure of such meteoric masses, which certainly never could be obtained by keeping it entire, merely as a museum curiosity, either here, in London or elsewhere.*[13]

For Selwyn, the 'very important results' leading from division clearly trumped the German's ostensible concern about the Bruce specimen's potential primacy among the British Museum's collection. By now Mueller would have been accustomed to such brickbats, even ones so adroitly put. Earlier, in yet another disheartening episode, the exploration committee of the Burke and Wills expedition did not pay due deference to his exploration experience. As president of the Philosophical Institute in 1859, Mueller was instrumental in its application for royal charter and was also in the chair in January 1860, when assent was granted for the title of Royal Society of Victoria to be used. He joined the expedition's exploration committee, the body responsible for overall management of the undertaking, as was appropriate for one of his experience and not merely because he was the society's president. In his role as general member, not office holder, he made valuable contributions to early arrangements. But his strong recommendation for expedition leader, the South Australian police commissioner and experienced explorer Peter Warburton, was overruled in favour of Burke.

The rebuff was a turning point for Mueller. He retreated from exploration committee business in general, unusual for one hitherto active in sub-committees and working parties. But he did provide thoughtful recommendations on equipment and scientific instructions when requested, even though the headstrong Burke largely ignored them. Mueller became pessimistic of the expedition's chances for success and took little part in its further activities. Planning for provisions, route, and communications would be left to others.

The snub must have hurt. With his extensive qualifications and field work experience, his Victorian excursions, and his participation in the Gregory expedition, Mueller was an authority on Australian exploration and had good reason to assume he would be treated as such.[14] He had already traversed the very Gulf country through which Burke and Wills would travel, yet the expedition leader nominated by the learned society of which Mueller was a founding member barely paid the botanist a courtesy visit, and never directly sought his guidance.

* * *

On 3 March 1862 the opinions of Foord and Neumayer, with Selwyn's and McCoy's affirmative contributions, were sent to Governor Barkly and James Bruce. In reply to Selwyn, No. 1's owner again declared his long-held view favouring division of his specimen, in terms of its general interest to science and the public, 'although I have always been inclined to modify my views in difference to higher scientific authority.' But Bruce was strongly opposed to what he called Neumayer's suggested 'mutilation.' Later in the month Mueller wrote to Owen advising him of No. 1's safe arrival in Melbourne from Cranbourne, and again wrote to Maskelyne, requesting his decision on the suggestion of exchange, since:

> *Several of the scientific gentlemen here regard strongly the division of this large and truly superb specimen. But on behalf of the British Museum I have entered a protest against such cause, until we have learnt your decision in reference to the proposed exchange.*[15]

The Royal Society's second annual conversazione was held the following month, in the evening of Monday, 28 April. On Victoria Street the society's hall was complemented with a large marquee at its entrance, in which refreshments were served, and the building's interior was bedecked with maps, drawings, models, fossils, and stuffed animal specimens. Many of the society's principals were in attendance, as was a recovered John King of Burke & Wills fame – the only survivor of the final exploration party – and Dost Mahomed, an Afghan cameleer who had been on the Cooper with the expedition.

Governor Barkly and party arrived and perused the exhibitions before he delivered his anniversary address. He noted the death of the Prince Consort, Albert, during the year and also that of society member Dr Ludwig Becker, another Burke & Wills casualty and an 'indefatigable contributor.' His summary of events surrounding the Cranbourne masses noted that the various questions on retention and division of the samples had been submitted to him. While taking credit for suggesting the exchange of a complete No. 1 for No. 2, Barkly acknowledged those opposing voices that sought a detailed analysis via a divided specimen. He professed pride that there were adequate men of science in the young country capable of carrying out an appropriate examination of the meteorite – 'this mysterious visitant of a bygone age' – and its characteristics. Then he took a step into Mueller's camp:

> *But I must confess, nevertheless, that looking to the uniqueness of the specimen … it strikes me it would be a pity to break it up before it has been seen by the scientific world, while the number of European savants is so far greater, that it would seem almost selfish to seek to anticipate on these distant shores their experiments upon it.*[16]

Barkly's position that the wider scientific community should see the unsullied meteorite seems more cosmetic than authentic. And the selfishness, or not, of anticipating future experiments repeats Mueller's limp assertion that No. 1's division would somehow degrade any subsequent analysis of the cut portions. In fact, samples were already being cut from meteorites large and small as part of regular scientific exchange and examination. The extracted pieces were considered representative of their source fragments. Barkly had himself engaged in the practice when he sent one of FitzGibbon's samples of No. 1 to the Imperial Museum of Mineralogy in Vienna. Those conversazione attendees with a mind to retention of their local artefacts, and particularly one so remarkable as the main Cranbourne mass, would have been quietly dismayed at the governor's stance.

So almost two years after Edmund FitzGibbon's paper on the meteorite's identification, the final placement of the specimens was still in question. The Abel mass was on display at the International Exhibition in London, after which it was likely to be sold to the British Museum. James Bruce's considerably larger sample was temporarily placed at the National Museum of Victoria, under the covetous watch of Frederick McCoy. Bruce now favoured a simple division and sharing of his specimen, but not its further 'mutilation.' Ferdinand Mueller was directly acting as an agent of Bruce, and of the British Museum, while being somewhat sympathetic to local concerns. Henry Barkly had similar empathy for his Royal Society colleagues but forwarded a potential solution to the British Museum that favoured the home institution. Alfred Selwyn and McCoy, with Georg Neumayer, Richard Daintree, George Foord, and others were agitating for retention of, at very least, one half of the Bruce meteorite. Their voices would grow stronger, and in the next few months, while a decision from London was awaited, another aspect would add to the fray – government intervention.

By Authority of the Commissioners.

Official Catalogue

OF THE

MELBOURNE EXHIBITION, 1854,

IN CONNEXION WITH THE

PARIS EXHIBITION, 1855.

"Come, bright Improvement, on the car of Time,
And rule the spacious world from clime to clime!
Thy handmaid Arts shall every wild explore,
Trace every wave, and culture every shore."

MELBOURNE:
PRINTED AND PUBLISHED FOR THE COMMISSIONERS,
BY F. SINNETT AND COMPANY.

Melbourne Exhibition Catalogue, 1854, where a 'Specimen of Iron from Western Port, and a horse-shoe made from it' were first documented. Publisher – F. Sinnett & Co., 1854. SLV

Edmund FitzGibbon, Melbourne Town Clerk, who identified the meteoritic nature of the Western Port irons. Sketch by Tom Roberts, *c.*1891. Pictures Collection, SLV

The Royal Society hall, Melbourne, scene of much debate over the meteorite's fate.
Royal Society of Victoria

Victorian Exhibition, 1861. Abel's No. 2 specimen was listed as 'of the greatest interest and of magnificent dimensions'. Photograph by Cox and Luckin, 1861. Pictures Collection, SLV

Georg Balthazar Neumayer. A pioneering contributor to Victorian science, he documented early findings on Cranbourne. Photograph by Frederick Frith, 1864. Pictures Collection, SLV

The Great Comet of 1858 over Melbourne. The late 1850s and early 1860s saw heightened cosmic activity visible from Victoria. Engraving by Ludwig Becker 1858. National Library of Australia

Ferdinand Mueller. Ingratiating, methodical, brilliant — he was tireless in pursuit of the main fragment. Photograph by J.W. Lindt. Pictures Collection, SLV

Frederick McCoy. The pugnacious leader of the push to keep the meteorite in Victoria, he fought a rear-guard action over four years. Photograph by Johnstone, O'Shannessy & Co., *c.*1870. Pictures Collection, SLV

Alfred Selwyn, Victorian government geologist. Not afraid to challenge authority, he was a primary organiser of the retention effort. Photograph by Johnstone, O'Shannessy & Co., *c.*1890. Pictures Collection, SLV

Cranbourne No. 1 on the day of its excavation in February 1862. Photograph by Richard Daintree, 1862. Pictures Collection, SLV

Nevil Story Maskelyne, Keeper of Minerals at the British Museum. He was greatly rankled by McCoy's prevarication. Photograph by William Henry Fox Talbot, c. 1844. MoPA

Victorian governor Sir Henry Barkly. An enthusiastic amateur scientist, he was a reluctant referee for the disputing parties. Engraving by Frederick Grosse, 1858. Pictures Collection, SLV

Above The Burke & Wills funeral procession, 21 January 1863. Charles Frederick Somerton 1863. Pictures Collection, SLV

Left Robert Brough Smyth, Secretary of Mines. He described the Cranbourne No. 1 fragment in the British Museum as 'an object of vulgar curiosity'. Drawing by George Gordon McCrae. National Library of Australia

John Macadam. The Scottish doctor and Royal Society member riled James Bruce with an 'insidious sneer'. Engraving by Frederick Grosse, 1865. Pictures Collection, SLV

The Director's Residence, Botanic Gardens, Melbourne – scene of the falling out between Ferdinand Mueller and Frederick McCoy. Henry William Mobsby, c.1920s. Mobsby Collection, Fryer Library, Brisbane

Sir Charles Darling, Henry Barkly's successor and third governor of Victoria. He issued the order for Cranbourne No. 1's transport to London. Photograph by Johnstone, O'Shannessy & Co., c.1865. Pictures Collection, SLV

Cranbourne No. 2. For many years known as 'Abel's Meteorite', it was spirited away by Frederick McCoy upon its return to Melbourne in December 1863. Source: Museums Victoria / Photographer: Rodney Start

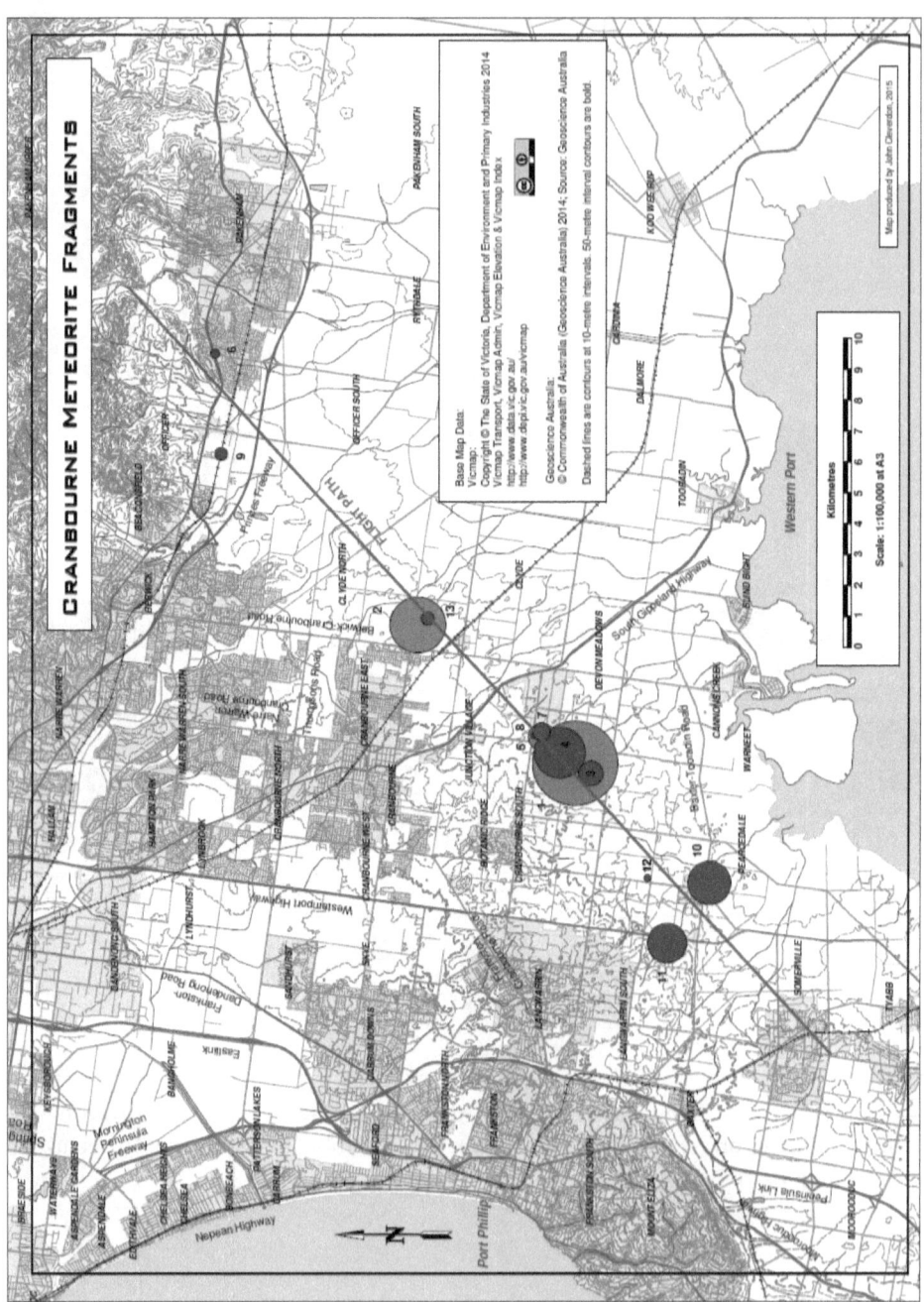

The Cranbourne strewn field, 22 km long. The direction of the meteorites' travel is north-east to south-west. Map by John Cleverdon, 2015. Used with permission

7
Of Mutual Interest and Support

*The vineyards of iron from where the meteor falls
are microscopic; smaller even than the fillings in our teeth.*
M.T.C. Cronin 'I don't know if he who is always waiting suffers more'

Meteorites are found all over the Earth (most rare are meteorites from Mars and from the Moon), but some locations have a higher concentration due to peculiarities of climate and geography. Finds are common in Antarctica, which acts like a continent-sized sheet of meteoritic flypaper; its cold and dry environment promotes preservation, and icefields afford a ready contrast for dark-coloured intruders. The icy whiteness of Greenland's backdrop helped reveal the Cape York specimens. A similar condition is found in the world's deserts, with sparse vegetation and uniform topography assisting in the identification of finds. In Australia, the prolonged aridity of the Nullarbor Plain and the Menindee Lakes area of western New South Wales provide optimal conditions for meteorite preservation and general ease of identification.[1] Proximity to a desert landscape may explain the discovery of meteoric iron in some Egyptian tombs; one, a find of decorative beads at Gerzeh, south of Cairo, and the other a ceremonial dagger in the haul retrieved from the resting place of Tutankhamun. Iron objects in Bronze Age Egypt, and with the nickel levels seen in these samples, could only derive from meteoric origins.

Iron meteorites can prove elusive even when evidence of their strike is overwhelming. Meteor Crater looks like a child's moonscape diorama, so evenly do its walls rise from the Arizona desert pan. But what of the meteorite that gouged this depression? It was mostly vaporised upon impact. With no comparable reference in our Earth-bound experience, it is difficult to comprehend the speed of cosmic bodies. Imagine the enormous kinetic energy required to directly transform an iron-nickel mass, dense metal approximately 50 metres in diameter, from solid to gas. The 30 tons of iron meteorites identified in the vicinity separated from the main body while high in the atmosphere, and take the name of the nearby community of Canyon Diablo.

The most common iron meteorites are the octahedrites, which when cut, polished, and etched with acid reveal a prominent linear matrix in composition, corresponding to an octahedron. Called the Widmanstatten pattern after the Austrian count who first worked on its formation, it is a crystalline encryption of the iron-nickel alloys kamacite and taenite. Found only in meteorites, these compounds comprise distinctive basket-weave ribbons, formed when their mass cools at an even rate over an extremely long time, in the order of millions of years.

Other external attributes characterise iron meteorites. Regmaglypts are the hollows, often likened to thumbprints, found in most meteorite types. Shaped by vortices of hot gases and the ablation of a meteor's surface as it passes through Earth's atmospheric rind, they are shallow features in the stony class. Among the irons, however, they forge deeper impressions and striking surface forms. Commonly presenting as dark and shiny laminar folds, their texture can resemble stiff peaks of whipped cream on a seemingly petrified dessert.

The International Exhibition of 1862 ran from May to November in South Kensington, London. In the decade since the hugely successful Great Exhibition of 1851, the 19th century appetite for world's fairs showed no sign of abating. Six million people passed through the doors of the exhibition hall, a cast-iron, brick, and glass construction which sprawled over 21 acres and housed a motherlode of natural wonders and human invention. The illustrated exhibition catalogue described Victoria's contribution, comprising 542 exhibitors, as 'an enormous collection … a more extensive and varied collection has never before been sent from any British colony to Europe.' Among its curiosities was a large gilded obelisk, representing the amount of gold found in the colony since 1851 – about eight hundred tons, valued at one hundred and three million pounds sterling. It was moved three times before the exhibition commissioners were satisfied with its placement, conspicuous on a platform beneath the building's eastern dome. Also on display was the quartz-crushing machine seen in operation at the Victorian Exhibition in Melbourne the previous year. It was awarded a medal in its Class 8 category, for 'efficiency, adaption, and practical utility.'[2]

Royal Society of Victoria members were well represented among the jurors' awards. Medals went to McCoy, 'for collection of fossils illustrating the order of the sedimentary rocks and for good engraved plates of fossils;' to Selwyn, 'for the progress of his survey alike important to the colony and to all geologists;' and to Daintree 'for photographs of rocks, fossils, and scenery, illustrative of Victorian geology.' George Rowe's eight watercolours of early Bendigo and surrounds were similarly recognised, 'for faithful and beautiful delineation of the country, workings, and other relations of the gold fields.' *Melbourne Punch's* exhibition correspondent, John George Knight, co-designer of Victoria's parliament house, reported in July that Professor Richard Owen had been by the Victorian collection and inquired about Cranbourne No. 2, only to be disappointed as it was not yet unpacked.[3] Despite Owen's curiosity Abel's meteorite made less of an impression than at its previous outing, being just one of many exhibits in the Victoria Department's large section of the floor plan, and with so much gold in view. Only an honourable mention was awarded, 'for the exhibition of a mass of meteoric iron of great size.' It must have been tough to compete with that obelisk.

In Melbourne, McCoy was entreating the Secretary of Mines, Robert Brough Smyth, for support. Smyth, a Northumbrian meteorologist, geologist, and mining engineer, was influential in all manner of Victoria's mining policy. He held connections with Adam Sedgwick and Sir Redmond Barry, and with McCoy, to whom he was commended by Sedgwick. Highly competent but officious, his demanding work practices polarised his subordinates and would bring him undone in later years. He was approached with a query by George Evans, namesake of the Sunbury squatter and Victorian postmaster-general in the current government under the premiership of John O'Shanassy. Evans wished to understand the circumstances of the meteorites with a view to retaining them in the colony.

Smyth wrote to McCoy on 15 April and explained that 'great anxiety was felt at the proposal of this national curiosity being removed from the colony and that the question was about to be raised in Parliament.' He asked for any information McCoy could provide on the subject, for Evans' edification. In mid-May McCoy, discerning potential allies to his cause, replied in detail with the sequence of events to date, reflecting on Bruce's change of mind and reserving some firm language for Mueller:

> *I trusted that especially as the matter was not in his department of colonial science that he would not urge any claims which he might have established upon it to the great injury of the Colonial Scientific Department to which it would properly belong even if he had perhaps hastily promised that it also as well as the first should go to the British Museum leaving no representative of so interesting a natural curiosity in the colony in which it fell.*[4]

McCoy was mildly appalled that Mueller seemed ready to allow both Cranbourne pieces to be acquired by the British Museum. In the Irishman's mind Mueller had 'used his persuasive powers with His Excellency the Governor who acquiesced in his view in total ignorance of the fact that I had been expecting myself to obtain the whole or a partial specimen for the National Museum.' Later in May McCoy, in response to another Smyth inquiry – regarding Bruce's potential asking price for his specimen – continued to publicise his displeasure with the botanist. He admonished Mueller for 'his exertions in the opposite direction' while there had remained a chance that Bruce would keep the meteorite in the colony. In this note McCoy seemed resigned to the suggested swap of the two main pieces. Curiously, over the exchange of these letters McCoy revealed that Richard Daintree was 'a personal friend of Mr Bruce the owner of the meteorite,' which makes sense of his earlier assignment of the young man to cajole Bruce but does not explain Daintree's failure to do so.

In July 1862 the Victorian parliament nominated a committee of inquiry under George Evans to consider and report on the envisioned removal of No. 1 from the colony. Appointees were Smyth, Selwyn and McCoy – despite the latter two's substantial involvement in the issue thus far. However, by the 28th Smyth was already writing to McCoy to inform him of a pause in committee proceedings before they had even begun:

> *I am directed to inform you that Dr Evans has postponed summoning the Board to meet in consequence of his intended absence from the colony for a short time and that he hopes nothing will be done with the meteor until he returns to Melbourne when he will cause the Board to be summoned without delay.*[5]

Evans had taken ship to Wellington, where he held interests in several properties. A classicist with an eloquent but monotonous oratory in parliament, he had earlier played a not insignificant role in the beginnings of colonial New Zealand, initially settled along Wakefieldian principles of systematic colonisation. Due to his weighty bearing and sonorous speaking manner he was a figure of some fun to his staff and colleagues in Melbourne. *Punch* had a great time with his public character; addressing him as 'this Methusuiean sage,' and 'the Founder of New Zealand,' even likening him to an Egyptian mummy in one if its satirical biographies. His departure across the Tasman suited McCoy, who appreciated the veneer of governmental protection afforded to No. 1 by the prorogued committee, even as the mass was ensconced in the museum he directed. In his early August reply to Smyth, McCoy was reassuring. He would 'take care that the wishes of the Minister shall be made known to any persons undertaking the removal of the meteorite and as far as may be permitted to me I will see that they are complied with.' Even so, Evan's absence was not the 'short time' Smyth envisaged; nearly a year would pass before the committee finally met and conducted hearings.

In the interim, the principals of the British Museum were finalising their strategy. Nevil Maskelyne wrote to Richard Owen on 21 May, suggesting that those in Melbourne who sought division of No. 1 were ignorant of the difficulties, in labour, expense, and time, of doing so. He considered an entire specimen more valuable than a fragment, and he would 'far rather have the small aerolite, than be a party to the division of the large one.' These benevolent utterings aside, he wanted Bruce's iron. Referring to Barkly's suggestion for a swap, he told Owen 'I think the proposition as originally made, a very good one, and as honourable to the patriotism as it is to the liberality and sagacity of the gentleman who proposed it.' He went on to explain his view of the order of things:

> *The fact is, the Colonial Museum has much to gain, by the sort of relations with us which the gentlemen at Melbourne are proposing in this case. It is, in fact, assigning to the Imperial Museum, and to the great Museum of the Colony, their proper relative functions of mutual interest and support.*[6]

Maskelyne proposed sending to Melbourne a selection of fossil casts and duplicate meteorite samples with No. 2, once purchased; an 'exchange' procedure being followed with museums in Calcutta and Madras around this time. In Maskelyne's mind 'they gain more than they lose, and so do we,' but it was a Hobson's choice. Certainly some casts and duplicates were no loss to his museum; these trades were heavily weighted in favour of the home institution. In this way the 'proper relative functions of mutual interest and support' were promulgated. It was 'best to do these things in regular order.' Owen was in complete agreement, writing to Murchison in that vein on the same day. Sir Roderick forwarded the two opinions to Sir Henry Barkly on 22 May, with a not-so-subtle reminder as to the honourable course of action:

> *I enclose the opinions of Professors Owen and Maskelyne, of the British Museum, on the subject of your great aerolite, and have to add that I fully coincide with them. It seems to be highly desirable to prevent the division of this splendid specimen, and I earnestly hope that your suggestions may be completely carried out, as being conducive to the advancement of science, the enrichment of the great British Museum, as well as highly honourable on the part of the colonists of Victoria.*[7]

September 1862 in Victoria heralded noteworthy terrestrial and celestial events. An earthquake shook the area north of Ballarat, with the town of Talbot having 'windows of several shops and private houses violently shaken, producing no small alarm among the occupants.' Further east, 'a tremulous motion' was felt in Taradale.[8] From the Williamstown Observatory Robert Ellery reported a comet in the northern sky, with the appearance of 'a moderately bright nebulous star' even among the strong moonlight of the month's first week. What would become famous as the Swift-Tuttle comet was also detected further north in New South Wales, from Windsor, by the ever-vigilant John Tebbutt.

At the Royal Society's ordinary meeting on Monday 8 September 1862, the topic of the Bruce meteorite returned to proceedings. Referencing his anniversary address earlier in the year, President Sir Henry Barkly spoke of the differing opinions on the meteorite's disposal and introduced the Murchison letters from London, read by society secretary Dr John Macadam. The *Argus* reported next day that the objections of Murchison, Maskelyne, and Owen to the division of No. 1 led to protests among 'several gentlemen,' including Georg Neumayer, who 'urged that the meteorite in question should be retained in the colony.' Here was the first bold statement from Melbourne's scientific community on their preference for placement of No. 1, although these views were not noted in the transactions for the meeting. What was entered to the record, however, was Edmund FitzGibbon's submission of a notice of motion for the next meeting:

> *That a communication be forwarded from this Society to the Executive Government, urging the expediency of retaining, if possible, in this colony the Bruce Meteorite, and of recovering the smaller Meteorite found in Victoria, and now in the International Exhibition, London; and suggesting that casts and specimens of both meteorites be presented to the British Museum.*[9]

Now things were out in the open. FitzGibbon and Neumayer had dropped the pretence of No. 1's division as a priority, and wanted the sample in its entirety. The suggestion of casts and specimens appears a mischievous reflection of Maskelyne's own proposal. Although McCoy's attendance at this meeting is not attested, his influence seems at work in that final phrase.

Barkly was uncomfortable with this turn of events. In a July letter to the Secretary of State for the Colonies, the Duke of Newcastle, he had enclosed his Presidential Address to the Royal Society in April. This was by way of explaining 'under what circumstances one of the largest meteorites in the world, conveyed at my expense to Melbourne, would be available for the British Museum.' Having suggested the exchange, he also asserted that 'no objection will now be raised to this course, and that so soon as the purchase of Mr Abel's meteorite shall have been completed by the Trustees, the larger one will be shipped home to them.'

Was Barkly a little embarrassed by the gold medal for Arts and Science sent to him via an earlier despatch from Newcastle, in May? It originated from the Emperor of Austria 'in consequence of my having forwarded to the Imperial Museum of Mineralogy at Vienna a portion of meteoric stone found in this colony.' After all, no portion of the meteorite had yet made its way to a British home institution. The governor had professed to 'being anxious at the same time that your Grace should not suppose that I have neglected the interests of my own country in the matter.' But now, far from 'no objection' being raised, he faced an insurgence among his Royal Society colleagues on the placement of the primary specimen. It was a vexing and painful issue.

Barkly was a skilful administrator, able to navigate the perils of Victoria's clamorous early efforts at responsible government. Among Britain's colonial possessions, none had experienced Victoria's whirlwind of excitement and rapid transformation in their first decade, the last years of which he had presided over. The diminution of responsibilities inherent in his largely symbolic role as constitutional sovereign was a departure from his West Indian experiences. Nevertheless, his governor's salary was the highest in the empire, due to the Colonial Office's estimation of the post's difficulty, a posture with some grounding in the travails of Charles La Trobe during the early gold rush. Victoria's runaway success and exalted international standing was a testament to the work of many years, and, of Barkly's two viceregal predecessors, it was the long-serving La Trobe who did most to lay stout foundations for the colony to come.

> Of French Huguenot descent, La Trobe was born in England and educated there and in Switzerland. He was a keen mountaineer who, by his early 20s, had made several ascents in the Swiss Alps. Also an ardent traveller, he wrote of his journeys as a self-styled 'rambler' in several books, including two that covered forays across the United States and Mexico in the mid-1830s. In October 1839, accompanied by his Swiss wife and young daughter, he arrived in Melbourne to succeed Lonsdale.
>
> Over the course of his tenure La Trobe would supervise Port Phillip's transformation from Van Demonian outpost to major British colony, and shepherd Melbourne through initial years of momentous change. But first, a hiccup – and a severe one. Melbourne land sales of the late 1830s

promoted intense speculation from certain quarters, with some investors making impressive profits on early purchases at later sales, often upon subdivision of the initial half-acre blocks. The Port Phillip banks provided willing lubricant for these transactions, but the beginnings of economic depression in Britain and the harsh drought of 1838–1840 drove wool prices down and reduced demand for stock in general. A cessation of land sales in Port Phillip followed, and with the decrease in land revenue governmental deposits were withdrawn from the banks, which began to call in loans. Early Melbourne was a strongly entrepreneurial community, and many were left with unsustainable financial burdens. The emergent economy contracted and a mini-depression in the early years of the decade drove insolvent debtors to bankruptcies and brought on much hardship. Several banks failed. Labour became scarce and expensive. Squatters, unable to sell their sheep or their wool, found a new market for their stock after Yass landholder Henry O'Brien popularised a process for boiling down animals for their tallow. Approximately half a sheep's weight could be rendered productively in this way. Boiling-works sprung up on the lower Yarra and Maribyrnong Rivers, a noxious but necessary alternative industry for a settlement that had thus far prospered off the sheep's back.

The district began its recovery in 1842, and as the decade progressed Melbourne embedded itself into its Yarra-side location. Ships jostled in Hobsons Bay, by the river's mouth, bringing settlers from Van Diemen's Land and, increasingly, directly from Britain. Overland traffic brought flocks and herds in great numbers. Pastoralist expansion spread north, west, and south-east. Residential housing and commercial enterprises in Hoddle's huge grid above the river fanned out from the south-western corner, over the Elizabeth Street gully and up into the manna gum forest to the east. The first bridge over the Yarra, an alternative to the rope-and-pulley punt arrangements used to that point, was built in 1845. But the river would not be tamed easily; serious floods caused disruption in 1842, 1844, and 1848, and a mighty tempest in November 1849 submerged the St Kilda Road and formed a lake that stretched from Emerald Hill to Flemington.[10]

Despite Governor Gipps's tight grip on the purse strings, the first civic and cultural institutions were established. In 1841 a post office was built in Collins Street, and a Supreme Court with resident judge was appointed. The first of several markets was set up, overseen by elected commissioners spread over four wards – a preface to local government. Creation of a jetty was underway at Williamstown and plans for a substantial gaol were developed. But the Sydney-based governor's economic caution meant delays in construction of a customs house, and when he disallowed repeated requests from La Trobe for funds to build a Mechanics' Institute, the officers of that embryonic organisation took matters into their own hands.[11] They built their own. Based on the British model, and ostensibly for the education of tradesmen, artisans, and working men, in time the Melbourne Mechanics Institute took the form of many of its Victorian contemporaries and became a hub for community amenity and activity. More than five hundred such buildings remain today.

In 1839 the first immigrant ship to sail directly to Port Phillip from Great Britain carried mostly Scottish 'assisted' immigrants. What had been a slow stream of arrivals in the late 1830s became a steady influx in the first years of the new decade. By mid-1841 Melbourne welcomed an average of one immigrant vessel per week, and by December 1842 Port Phillip's population was close to 24,000. But the withdrawal of the government subsidy after 1842, caused by the economic downturn, effectively stopped overseas immigration. It would not resume in earnest until 1848.

Arrivals to Port Phillip joined a population of domestic servants and agricultural labourers, merchants and traders, pastoralists and planters. The growing town on the Yarra was recognised in 1842 when the New South Wales Gazette published the Melbourne Incorporation Act, and after elections in December Henry Condell became Melbourne's first mayor. In the following year the Port Phillip District sent six representatives to the newly formed New South Wales Legislative Council.

Significant building works were gotten underway in 1846, with foundation stones laid for both a new Yarra bridge and Melbourne hospital on a sunny Friday in March. A grand procession marked the dual occasion, with the town's various teetotal societies and lodge members turning out in their finery to form a picturesque march along Swanston and Collins streets. La Trobe officiated at the bridge's initiation, and Mayor Condell at the hospital site. There, a Latin-inscribed vellum scroll and some coins were placed under the foundation stone, to be revealed to future generations. The next year Melbourne was proclaimed a city, Queen Victoria's Letters Patent also conferring status as a cathedral city, with Anglican and Catholic dioceses founded. Gas lighting illuminated some shops, driven by primitive coal gas apparatus, and by 1849 the main streets were grubbed of their tree stumps and paved.

As the district grew in population, so it did in confidence. An independent streak was apparent among Port Phillip settlers from the late 1830s, driven by resentment at Sydney's appropriation of local revenues and a lack of adequate representation within the legislative council of that distant government. Separation was mooted as early as 1836, and a robust Separation Association actively pursued its agenda via petitions to Gipps in 1841 and 1845. In 1848 the cautious La Trobe thought it 'very unlikely that separation will take place in a hurry'[12] and trusted in the outcome of constitutional reforms planned by the Colonial Secretary, Earl Grey, to address the issue. La Trobe's risk-averse attitude won him few friends among the town council and further provoked his unpopularity across the general population. The *Argus*, an unrelenting critic, derisively referred to him as 'the Hat and Feathers' as if his superintendent's trappings were the limits of his substance. The newspaper's proprietor and editor, the remorseless Edward Wilson, branded La Trobe 'a tool of the squatters' and accused him of following 'a sneaking, treacherous course.'[13] Tall poppies have long been targets.

But change was on the way, and the Australian Colonies Government Act of 1850 finally conferred colony status upon the Port Phillip District, to be named Victoria, and signalled the beginning of responsible government – to take effect from July 1st, 1851. The district could look forward to a bright future, with the flourishing port of Melbourne as its comfortable capital, 'the faithful servant of a prosperous and compact hinterland.'[14] But none expected that the next few years would see a spectacular rise in Victoria's fortunes, indeed for it to be ranked among the richest territories in the world. La Trobe's classics education ensured his familiarity with Augustus's boast of Rome; that he 'found it built of brick and left it clothed in marble.' The rambling mountain climber would withdraw before the completion of Melbourne's transformation from wattle-and-daub settlement of 2,000 souls to gold-appointed metropolis of over 75,000 population. However, his stewardship of the town and surrounding district through the depression of the early 1840s, and subsequent recovery, meant the young colony had leapt its first hurdle. Unbeknown to its citizens and government, Victoria was poised for great change and its greatest challenge: it would soon be celebrated as a southern El Dorado.

8
This Out-of-the-way Part of the World

You are speeding in thin sub-zero air
over some artist's palette, of white salt
and red ferric rocks rusted
with oxygen, heat and time,
the cracked cooled scum of the molten-iron planet.
Mark O'Connor 'Dot Paintings'

The Moon's lack of crustal movement and a tenuous atmosphere, stripped over time by the caustic solar wind, means its impact evidence remains forever stationary and visible. On Earth, however, the development of a crust, with continent-sized plates actively spreading and grinding, has worked to conceal such traces. And the formation of the atmosphere and oceans, and weather-driven erosion, means the signs of the planet's scourging are often subtle.

Cannonades of space-borne grapeshot abraded Australia's dermis over immense periods. But you still need to know where to look for marks of this molestation. Impactor velocity is nigh on incomprehensible, and their strikes freighted such momentum that they churned the Earth's rocky crust like butter and set fire to the world. But you still need to know how to look for craters. Invisible to the eye and even satellite imagery, some impact sites are so ancient they are now subterranean relics of former surface formations, entombed by time and relentless corrosion, and may only be divined by the penetrating inquiry of geologists' electronic instruments.

And so, evidence of asteroid-size collisions is often fossilised deep underground, such as at the wonderfully titled Tookoonooka crater in south-west Queensland. At 120 million years old, it makes a possible pair with another Queensland crater, Talundilly, in evidence of a binary asteroid impact. Woodleigh dome, by Shark Bay in Australia's far west – an ancient locale, scene of rock formations and early-life structures – is a much older concealed feature, dating from 365 million years ago. But the Acraman crater in the Gawler Ranges of South Australia, even more ancient at 580 million years, is pleasingly visible in the form of a roughly circular lake, 20 kilometres in diameter, which occupies the middle of a much larger depression.

The great Australian craters, subterranean or exposed, occupy a wide swathe of the continent from Western Australia, east through the dry country of South Australia and the Northern Territory, and peter out into Queensland's outback. The younger examples, less weathered, are readily recognisable. Wolfe Creek's walled circle, thrown up in the Pleistocene, seems the epitome of an impact crater – a smaller and shallower Barringer in the Kimberley. The Henbury site south-west of Alice Springs is a cluster of obviously meteoritic impacts, the result of an iron-nickel body fragmenting high in the atmosphere before collision. This is an unusual strewn area. Like a barrow field in concave, it has a dimpling of craters, rather than scattered meteorites, as its main feature. One of these hollows displays a rayed ejecta field, a formation seen in lunar and Martian craters but visible almost nowhere else on Earth.

Edmund FitzGibbon's September motion, recommending governmental assistance in retaining No. 1, was not actually brought forward until the Royal Society's meeting of 17 November 1862, in what was a pivotal gathering. Sir Henry Barkly was not present, having promised his attendance at a lecture by John Fawkner ('Reminiscences of the Colony') at the nearby Princess's Theatre. In his absence Ferdinand Mueller took the chair. After initial business FitzGibbon moved his proposition, although with some reluctance. In the two months since his initial submission there had been commentary about his reasons for making it, and perhaps not all was complimentary. FitzGibbon cited public grounds for his motion, and that he 'did not wish to be considered as having any selfish interest in the matter, having only in view the furtherance of science.'[1] He gave a full account of the meteorite's discovery and ensuing events, and mentioned the main objections voiced against his proposal, i.e. a breach of faith with the British Museum and the general interests of science being better served by sending No. 1 away.

FitzGibbon addressed both points. He argued that James Bruce had made no 'binding promise … to present the meteorite, or a section of it, to the British Museum.' As for the interests of science, the sending of a fragment, a section, and a cast would aid just as well as the meteorite in its entirety. 'The cause of science would be better served by leaving such an object of interest in the neighbourhood of which it was found rather than sending it away to enrich some already crowded collection.'[2] Robert Brough Smyth seconded the motion, describing the certain destruction of the sample's magnetic properties if it was removed, and emphasizing its pre-eminent position among the world's meteorites in size, value, and reputation.

Mueller heard them out, then played his trump card. He opposed the motion, stating that the meteorite had indeed been promised to the British Museum by James Bruce, and that he was acting as Bruce's agent and was trustee of the specimen. He read from Bruce's letter to him where the owner stated that he hoped Mueller 'will leave no stone unturned to get it placed within the walls of the British Museum.' Hugh McKay provided additional ballast via a document, also put forward, wherein the farmer stated that 'he only sold the meteorite with the object of placing it in the British Museum.'[3]

This seemed to settle the matter. Alexander Smith, the Scottish engineer behind Melbourne's gasworks, certainly thought the issue closed and recommended moving on to the other business of the evening. Surgeon William Gillbee spoke to the crux of the problem, saying 'the meteorite was not the property of the society, and they therefore had nothing whatever to do with it.' Other members were less adamant. John Macadam recognised a promise had been made but regretted that such a valuable specimen would be lost to the colony. He questioned James Bruce's scientific 'attainments' and contrasted him unfavourably with a namesake from the Victorian railways, considering that 'if that gentleman had owned it he would not have sent it out of the colony.'[4]

William Crooke, seeing the strength of Mueller's position, moved an amendment to FitzGibbon's motion. He proposed that a committee be appointed 'to take such measures as may be deemed best calculated for securing possession of the Bruce Meteorite for the Colony of Victoria.' FitzGibbon, accepting the meteorite as Bruce's property, withdrew his motion and adopted Crooke's amendment. It was passed nine to three.

But if Nevil Maskelyne and Richard Owen of the British Museum looked likely to receive Bruce's donative, they were left in no doubt about the £300 price tag accompanying Abel's specimen – his brother, Johann Leopold Abel, had been busy reminding them. The father of Frederick, Augustus Abel's nephew who had earlier summarised his uncle's character to Maskelyne, Johann was a pianist and composer, and a former child prodigy. He appeared to be now acting as Augustus's agent in London, not Frederick, and had met Maskelyne at the British Museum in the late summer of 1862. Summarising their interview in a following letter dated 8 August, Johann understood from Maskelyne that his brother's meteorite 'has under your direction been actually bought for the British Museum, with the sanction of the Trustees of that institution under a special grant from Parliament.'

Not quite, it seemed. Maskelyne's written reply the next day indicated that approval for the purchase was still pending, and in fact he and Owen had not yet proposed the transaction to the museum trustees but would do so at their next meeting, in October. His bullish position in the meeting with Johann Abel may have been a defensive move, to ensure his vendor did not go elsewhere. Assuring the German that the recommendation would be adopted, he requested Abel's invoice. Johann obliged, without rancour, and by later September viewed the matter 'as about closed.' However mid-November found him still chasing Maskelyne for news of the trustees' decision. Payment was not authorised until April the next year.

* * *

Both the *Age* and the *Argus* reported in detail on the Royal Society's 17 November meeting, and now the Melbourne newspapers became the forum for argument. This print-version town square made space for observers, commentators, and participants alike. A scribe with the pseudonym 'Monomaniac' contended the meteorites were not space-borne at all, but rather the result of volcanic activity, and would lead to the discovery of other 'similar masses ... the value of the prize will more than repay, both in utility and value, the search for gold.'[5] Engineer Charles Mayes had a revelation after attending the society's November meeting. He wrote to the *Argus's* editor to assert that 'in all Crown grants made to purchasers of land from the Government of Victoria, the minerals are reserved as the property of the Crown; and consequently that unless the person who removed the Bruce meteorite previously obtained a licence to dig, search for, or remove minerals from the Cranbourne district, the meteorite is still the property of the Crown, and can therefore be disposed of as the Government think fit.'[6] *Melbourne Punch*, in its expose of events of the 17th, took aim at the 'frightful prolixity' of that evening's discussion, teasing Mueller for his approach and making light of suggestions to 'exchange our local prodigy for a score or two of trumpery casts which any Italian image vendor would supply us with for a few pounds.'

James Bruce failed to see any humour in the issue. On 3 December 1862, he penned a letter to the editor of the *Argus* 'to put the public in possession of the real facts of the case.' He explained his arrival in Cranbourne and purchase of McKay's iron mass for a nominal sum 'on the understanding that it was to be sent to the British Museum.' The early offers from FitzGibbon and McCoy, in August 1861 and January 1862 respectively, were discussed, including the crucial conditional proposal to McCoy of half the sample, let lapse by the Irishman's unhurried response. On 31 January 1862, Bruce had 'finally handed it over to Dr Mueller for presentation to the British Museum, according to my original intention.' He declined Selwyn's March overture as he 'considered I was bound in my promise to Dr Mueller.' However, if Mueller consented to the specimen's division 'I should not stand in the way.' Bruce went on:

> *To meet conflicting interests, and for the sake of an amicable settlement, it was after-wards agreed that it should be left to the authorities of the British Museum to decide whether they should*

> *purchase Mr. Abel's meteorite, and send it here in return for the larger one, or have the latter divided. Why should this arrangement be repudiated now? I have gone this much into detail because Mr. Fitzgibbon stated there was no binding contract on my part; probably he meant there was no formal legal document.*

He tried hard to keep his anger in check as he worked 'to prevent any future misrepresentations.' But, riled by comments made at the society's recent meeting, his restraint did not stretch as far as Macadam:

> *As for Dr Macadam's insidious sneer with respect to my 'scientific attainments,' they may or they may not be empirical; at all events, I have not thrust myself before the public. If the great doctor's last lecture is a fair specimen of his scientific attainments, I scarcely think he is free from the taint. But this is beside the question, I have yet to learn that, unless I am possessed of great scientific attainments, I cannot deal with any property I may have possessing a scientific interest, as I see fit, without consulting even the Royal Society.*

Bruce took umbrage with Macadam's perceived self-promotion as well as his verbal slight. He continued in high dudgeon:

> *Let the doctor commence to weed nearer home; there is plenty of room for the knife. I have lived long enough to know that they are not the men of greatest scientific attainments who are continually thrusting themselves before the public.*[7]

The meteorite's owner composed a reasonable slap-down. Macadam's response, if he made one, is not recorded. He was a high-achiever, even in a time when multiple talents seemed commonly visited upon determined individuals. A chemist, academic, politician, and non-practising medical doctor, he was active in Victoria since his arrival from Scotland in 1855 and accomplished a good deal during his ten years in the colony. Appointed to teach at Scotch College prior to departing Britain, he also lectured at Geelong College's antecedent institution and held the position of Victoria's analytical chemist. From 1860 he was Melbourne's health officer. Instrumental in the foundation of the University of Melbourne's medical school, in 1862 he became the faculty's first lecturer.

Macadam left a legacy of academic and governmental achievement, and he also entered Australia's cultural lexicon when his friend Ferdinand Mueller named the genus of a nut-bearing native tree after him. This was later exported to Hawaii and South Africa, and over 200,000 tonnes of Macadamia nuts are produced globally today. Macadam even had a hand in the early history of Australia's indigenous football code, when he co-umpired a pivotal match in 1858 between two Melbourne school teams, of a format considered a predecessor to the modern Australian rules football.

If Macadam did not reply publicly to Bruce's slights, FitzGibbon was less reticent. On 6 December he wrote to the *Argus* (although this letter was not published until 27 December), claiming 'Mr Bruce attributes to me a statement before the Royal Society which I did not make, and I cannot otherwise so effectually correct his own error, and that into which he misleads others.' He was referring to Bruce's assertion that 'Mr FitzGibbon stated there was no binding contract on my part; probably he meant there was no legal document.'

It was a lengthy rebuttal. FitzGibbon began by re-stating the remarks that prefaced his motion of 17 November meeting:

> ... *That if it should be shown that any engagement, pledging public or private faith, had been entered into with the authorities of the British Museum for the fulfilment of Mr. Bruce's declared intention respecting the meteorite, I should at once withdraw my motion.*[8]

The town clerk reasoned that because Bruce had offered one half of his sample to McCoy, he could be assumed to have a 'disposing power' over the other half. That being the case, a proposition by the Victorian government (which had already appointed a board on the subject) to Bruce 'would induce him to abandon his original intention in favour of the proposition which I then submitted,' i.e. the swap of specimens.

His positioning statements delivered, FitzGibbon proceeded to argue the case for retention with a series of well-crafted opinions, perhaps recalling his legal training: his arrangement would serve just as well as sending the big block in exchange for the small, the voyages would be risky, the meteorites were 'incapable of insurance.' More broadly, Melbourne would become an attraction for men of science, and the city's colonial youth would receive 'a taste and inclination for scientific observation and inquiry.' In a notable passage he played down the common assumption that the British capital was the premier location for such a scientific paragon, and suggested that Victoria might yet throw up a Benjamin Franklin or George Stephenson:

> *That it is by no means certain that scientific inquiry in respect to it may not be as effectively pursued in Melbourne as in London, since the annals of science show how that of electricity is indebted to a hardworking colonial tradesman, and how the locomotive steam-engine had its beginning, not in the hall of a museum or learned society, but in a wretched north-country cottage.*[9]

FitzGibbon reinforced that upon Bruce's letter being presented by Mueller, he had withdrawn his motion and voted for Crooke's amendment. In his view 'there has clearly been no denial of Mr Bruce's right to dispose of his property as he should think fit, but a recognition of that right and a desire to appeal to his judgement in the exercise of it, through the medium of the Government board appointed, with respect to these specimens.' He closed with a firm renunciation, and repeated his grounds for assuming Bruce would be open to the suggestion of keeping No. 1 in the colony:

> *Mr Bruce has, therefore, no ground for the resentment which he appears to feel at the supposed attempt to invade his right, or at the fancied repudiation of his engagements. It is evident that he did not consider his declaration to me of his intention to send the block to the British Museum as precluding him from promising one half of it to Professor M'Coy, nor his subsequent repetition of that declaration and delivery of the block to Dr. Mueller, as preventing him from again promising half of it to Mr. Selwyn, subject to Dr. Mueller's concurrence. It was not unreasonable to suppose that his power of disposal extended to the whole as well as to the part; and an appeal to his judgement on behalf of our own colony ought not to be considered as an insult.*[10]

Finally, FitzGibbon displayed the puzzlement of one possessed of a firm local outlook, but confronted with an opposing mindset and imperially focused point of view:

> *His desire to repay the pleasure and profit which he derived from the British Museum is worthy and laudable, but he must be a rare exception from the most of us if he do not confess that his gratitude should be still greater to the land we live in.*[11]

Bruce was indeed an exception to FitzGibbon's rule, as he noted in his reply of 30 December:

> *I must confess I am that rara avis to which Mr. Fitzgibbon alludes. His views with respect to this meteorite are connected with country, while I take a world-wide one; and I think this unique specimen will be visited by more and rarer men of science if placed in the British Museum than if buried in this out-of-the-way part of the world.*[12]

And there is the nub of the issue, articulated by its main protagonist. In Bruce's opinion his international view, buttressed as it was by Mueller's and Barkly's, trumped the parochial focus of FitzGibbon and his like-minded colleagues. But his converse preference for the meteorite's placement only ever recommended a single repository, the British Museum. Hardly an 'international' view, particularly when combined with the dismissive remarks about his temporary home. As it turned out, these December letters were Bruce's parting shots. In early 1863 he quit Victoria and returned to Britain, leaving an agent to lease his Cranbourne property and Mueller to manage affairs surrounding his windfall meteorite.

* * *

Over this summer period the unfortunate Burke and Wills once again vied for Victoria's attention. In the days before Christmas, more than a year since the colony learned of their fate, their bones arrived by ship from Adelaide in the charge of bushman and explorer Alfred Howitt. Howitt had earlier been sent north and recovered lone survivor John King, interring the found remains of his lamented leaders. In this second recovery mission, again without fuss or loss of personnel and equipment, he unearthed the previously buried bones from the distant Cooper and transported them home. The remains lay in state at the Royal Society's hall for two weeks while a procession of 100,000 people viewed these relics of the fallen heroes.

Unbeknown to the expedition's late leaders, in their final days they had scrabbled over the eastern shoulder of an immense geological subsidence, later theorised to be one of the largest asteroid craters in the world. The Warburton Basin is thought to be caused by a strike over 300 million years ago, in which the parent asteroid – up to 20 km wide – split in two before impact. The resulting craters, 200 km in diameter, have been buried deep by geological action but surface remnants manifest in the broad depression, through which Cooper Creek flows fitfully and where Lake Eyre persists a long cycle of sporadic inundation. It was a suitably epic mise-en-scene for such a tragic opera.

On 21 January 1863, approximately 40,000 attended Australia's first state funeral and saw the explorers laid to rest in the Melbourne Cemetery, half a mile from the scene of their mildly chaotic departure two and a half years before. The *Argus* reported the event in detail, including the impressive size of the funeral procession – it stretched the length of Bourke Street from Spring to Elizabeth streets – and the many dignitaries present. It noted 'In Wills's coffin was placed the wreath his fellow labourers at the Magnetic Observatory had hung upon the shell, and upon both his and Burke's, laurel wreaths sent by Dr Mueller were deposited.' It was a magnanimous gesture from Mueller, who had advocated strongly for a different candidate than Burke to lead the expedition. Furthermore, the society's Exploration Committee overlooked, or ignored, his recommendations on equipment and supplies, drawn from his own experience as a solo explorer and participant in Gregory's North Australian expedition. Now a committee of a different stripe was about to serve him a more direct rebuke.

9
So Magnificent a Cabinet-Piece

Little enigmas strewn
like fossil raindrops over
the Nullarbor Plain.
Jan Owen 'Tektites'

The hyper-velocity collisions that create impact craters also generate an interesting by-product. A splash of metamorphosed terrestrial material and liquified ejecta is usually flung out from the perimeter of a major strike, and tektites are the small, often glassy, remnants of such material. They form a curious collection: some like knucklebone toys, others with coin shapes and raised flanges. Charles Darwin was the first to describe an Australian tektite, given to him by Thomas Mitchell, in 1844. Tektite variations arise from their formation. With liquified rock as their starting substrate, they rapidly cool upon rotation through the air, and some return to the ground as signature teardrop and dumbbell shapes.

Tektites are grouped into their own strewn fields, four of which cover the Earth. The Australasian field contains dark-coloured specimens from Southeast Asia, the Philippines and Australia. Evidence suggests they relate to an 800-million-year-old impact crater on the Bolaven Plateau in southern Laos.

Tektites belong to a group named impactites, terrestrial rocks deformed under the great heat and pressure of a cosmic blow. Other impactite examples are glassy forms that contain sand from the target rock field. In the rugged western ranges of Tasmania is a circular rimless basin, hypothesised to be an impact location. Concealed within mountainous terrain, it was a dusting of varicoloured glassy shards across the surrounding landscape – called Darwin glass after a nearby mountain, itself named after the famous naturalist – that suggested an impact source in the vicinity. And sure enough, in 1972 geologist Ramsay Ford identified what would come to be called the Darwin Crater.

Complex craters are formed when adequate impactor mass and velocity, and a near-right angle of impact, imparts such kinetic energy that intense shock waves are generated in the target rocks, permanently deforming them, and a sizable crater is excavated. Extreme pressures and temperatures obliterate the projectile. Then, very quickly, an uplift form is created by the rising of lower-down target rocks at the centre of the contact, and rocks at the perimeter collapse inward. It's like dropping a pebble into a pond, in that the violence of the action is so acute that crustal rock temporarily takes on liquid characteristics and a dollop of breccia will settle in the middle of a wider circular excavation.[1]

Gosse's Bluff, sited in the Northern Territory and neighbouring the marvellous Uluru and Kata Tjuta sandstone formations, seems a relatively small depression, five kilometres in diameter, but it is an illusion. The structure, known as Tnorala to the indigenous Western Arrente, is actually the remains of just the central uplift portion of the original, now eroded, crater, thought to be some 22 kilometres in diameter.

In January 1863 Nevil Maskelyne wrote to Sir Henry Barkly informing him that the British Museum's trustees had now secured Abel's meteorite, and requesting that he 'give the requisite notice to forward the larger Bruce Meteorite in exchange for it.' The governor was unsure how to reply, given the rumpus around the meteorite. He wrote to McCoy on 16 April:

> *I do not exactly know how to proceed in the matter, but as you are the actual custodian ..., I apply to you in the first instance to know how you are prepared to act. The Colonial Office have sent out a circular despatch stating the desire of the Imperial Government that specimens of the kind wherever procurable should be obtained for the British Museum.*[2]

Barkly enclosed the despatch with his letter, saying the documents were 'very much at your service' – a gentle reminder for the museum director of the home government's designs on the meteorite housed in his institution.

Barkly's letter and its enclosed request from Maskelyne re-animated McCoy, uncharacteristically quiet during the noteworthy events of the summer. Despite Mueller's tactical victory of 17 November, mitigated as it was by FitzGibbon's vigorous newspaper defence and the society's set up of a new committee, he sought out George Evans, now returned from his New Zealand secondment, and prompted the activation of the Board of Inquiry's hearings. McCoy referenced his recent discussion with Barkly. 'His Excellency quite agreed with me that the proper course would bring the matter under his notice so that official sanction should be given to whatever course was determined upon in relation to the disposal of the specimen,' he wrote on 21 April. 'I promised if possible to communicate with His Excellency on the subject on Thursday or in time for the mail.'

On the same day, a Tuesday, Barkly followed up his letter of the 16th with another to McCoy, suggesting the professor visit him in the afternoon of the coming Thursday 23rd to discuss the meteorites. The governor requested he consult with Mueller beforehand, because he wanted a document from McCoy that he could send to Maskelyne 'as a reply to his request that the arrangements entered into between Dr Mueller and yourself should be carried out without further delay.' The next mail departure was due on Thursday 23rd.

But McCoy hedged. He did not make the meeting, believing the Governor was not in attendance as the official flag was not hoisted when he called (it was a holiday; Saint George's Day). He did not reply to Barkly until Saturday 25 April, and used the opportunity to impugn Mueller's actions yet again:

> *The Director of the Botanic Gardens having taken upon himself to present the meteorite to the home museum without consulting the Victorian Museum Department, or the Government, excited I am informed considerable public indignation and (without any action on my part) the matter was brought under the notice of the Government some nine months ago, and the Honourable the Chief Secretary appointed a Board with a Cabinet Minister as Chairman to consider and report on the proposition for depriving the Colony of the valuable specimen and I as Director of the Museum received a letter directing that no steps should be taken for its removal without the knowledge of the Government.*[3]

The ploy worked. In a note to McCoy the next day, Barkly, only now aware of the existence of the Board of Inquiry, wrote:

> *I am sorry not to have seen you about the meteorites, but think on the whole I had better not acknowledge Mr Maskelyne's letter by this mail, but wait till I get the formal report of the Board and also see Dr Mueller as Mr Bruce's representative in this country.*[4]

But the Governor made clear his view of division:

> *I confess I am sorry that the idea of cutting the Bruce meteorite in two is to be invited on, as in my humble opinion it is a barbarism which will not tend to raise the character of the Colony.*[5]

McCoy's prevarication ensured Barkly missed the mail, and that Maskelyne and Mueller would be stalled a little longer. By the early 1860s marine steam technology and development of the screw propeller had improved the speed and reliability of mail transport between Britain and Australia. Sailing ships had traditionally used the clipper route and leveraged the strong westerlies of the roaring forties to cross the Indian Ocean from the Cape of Good Hope, on occasion taking a great circle route through even higher latitudes to shorten travel times. Steamships, not beholden to wind, could take a shorter course through the Mediterranean to an Egyptian port. After cargo and passengers were overlanded to Suez at the head of the Red Sea, another steam-powered vessel made for India or Ceylon via Aden, then on to south-western Australia and eventually ports on the east of the continent. Consideration was given to how much coal could be carried at the expense of cargo, and the distance between coaling stations was a factor in these journeys.

Steamers of the royally chartered and impressively titled Peninsular and Oriental Steam Navigation Company, later to be P&O, carried post under contract. Ships journeyed between Melbourne and King George's Sound in Western Australia, sometimes with a stop at Kangaroo Island, then to Point de Galle in Ceylon. From there, vessels of the India Line carried on to the Gulf of Aden and up the Red Sea to Suez, with mail next transiting Egypt by railway to Alexandria before the opening of the canal in 1869. From this port, ships for Marseilles took letters and packages to be overlanded from the south of France to London, or Southampton-bound packets made the final seagoing leg via Malta and Gibraltar. Five weeks was the average elapsed time from end to end, with trips of a four-week duration occasionally recorded. It was a marked improvement from the lengthy intervals and general uncertainty of the sailing era, but correspondents in Australia were still understandably keen to get their letters on the scheduled outbound boats. To 'miss the mail' was to add another month to one's wait for a reply.

The Board of Inquiry finally met for the first time on 24 April. McCoy's letter of the 25th to Barkly foreshadowed its interim report, stating 'the Government informs me of their desire to retain the half of the meteorite presented to them by Mr Bruce and to respect the right reserved for the British Museum by that gentleman of cutting off half at their own expense if they saw no impropriety in it.' He mentioned the 'principal scientific men of the Colony' in arguing for retention, and for additional clout referenced the findings of Austrian physicist and mineralogist Wilhelm Haidinger, who had published papers of his examination of a sample of No. 1, the same piece sent earlier by Barkly to the Emperor.

McCoy's hand is evident in the board's draft report delivered to the Chief Secretary (the final document cannot be located). The now-familiar positioning regarding Daintree's involvement and McCoy's initial overtures to Bruce are there, and the all-important offer by the farmer to divide his sample between the British Museum and the National Museum of Victoria.

Among the board's recommendations was that 'they be authorised to communicate with the Trustees of the British Museum with the view of retaining, if possible, the large meteorite for the National Collection of the country.' It held the opinion that 'the best interests of science' would not be served by the meteorite's removal to the northern hemisphere because this would leave incomplete the magnetic observations begun by Neumayer. A curious observation, since the young German had previously recommended to Selwyn that the mass be sectioned to provide samples for exchange with other museums, an action sure to alter its magnetic character – if the weathering process of many years had not already 'overprinted' its primary magnetic record with terrestrial characteristics. In fact the board found, 'there are indeed, no scientific reasons for sending it home, and the risk involved in sending this, and returning the smaller one, across so great an extent of ocean is in itself a good arrangement against the exchange proposed.' Insurance would be inadequate, and the loss of the specimen would be not only 'injurious to science,' but make Victoria a laughing stock; 'scientific men would look with wonder and pity on the people who could so lightly part with such a treasure.'[6]

But the report's strongest language was reserved for Mueller, singled out for apparently misleading Bruce on the level of agreement between him and McCoy, and for stepping on McCoy's patch. 'Mr Bruce, as he stated in a letter to Professor McCoy, believing that Dr Mueller was acting with the knowledge and consent of the Director of the National Museum (within whose promise it is to effect exchange of minerals with other public institutions) agreed to the proposition thus made.' And the report's conclusion had another reprimand for the botanist:

> *The Board desires to express their strong disapprobation of the conduct of the gentlemen who have, in any way, been instrumental in urging Mr Bruce to accede to the removal of the meteorites, who it appears acted all through as if desirous of being guided by the advice of scientific men. The Board can only report that the advice was not dictated by more patriotic virtues.*[7]

Mueller must have felt keenly mistreated by the board's draft findings. He had wished to appear before it and give evidence, but a mixture of circumstance, mis-communication, and McCoy's deft manoeuvring thwarted him. On 22 May Barkly wrote to Mueller and informed him of a surprising communication received from Smyth in relation to the board's activities:

> *I learnt yesterday that the Board had sent their Report to the Chief Secretary, and that you had not 'applied' to be heard! As I sent your letter claiming the right to be heard to one Member of the Board, and pointed out to a second Member the injustice of casting reflections on your conduct in this matter, without affording you an opportunity of justifying yourself, I am certainly astonished at the ground assigned for not examining you.*[8]

Barkly's astonishment was well founded, since he had engaged McCoy over Mueller's application, and also pleaded the case with one of Selwyn, Smyth, and Evans. He closed in some exasperation:

> *I shall write to Professor Maskelyne therefore referring him to the Board, and shall wash my hands of a matter which has occasioned me more real pain than anything that has occurred since I came to the Colony, because I feel that the course which has been indicated by the Board will place me in a false position, and throw discredit on all concerned.*[9]

Barkly's epistolary hand-wringing continued in his despatch to Maskelyne three days later. Referring to Bruce's *Argus* letter of December 1862, where the owner described his January suggestion of division to McCoy, he confessed ignorance of 'any such alteration of Mr Bruce's views' before, as he had told Mueller, abjuring responsibility for the affair:

> *... But I feel that it is very difficult for me to stay silent in the matter, and as the Board go on as to request that they should 'be authorised to communicate with Mr Bruce and the Trustees of the British Museum with the view of returning if possible the meteorite in the National Collection of that country,' it appears to me better that the future conduct of it should be left in their hands.*[10]

Realising he had overstepped in his July assertion to Newcastle that no objection would be raised to the course of proposed exchange, Barkly closed with an explanation, and a subtle deflection in McCoy's direction:

> *In alluding to the proposed exchange in my Address to the Royal Society of Victoria in April 1862 (which is I presume the ground on which you have been written to me on the subject) I was under the full impression that no objection could possibly be raised to a course which I learned had been suggested by Professor McCoy, even before I threw out the idea, as the best mode of settling the difficulty.*[11]

Maskelyne, now without the ongoing assistance of Barkly, would not be deterred by 'two or three persons connected with the museum at Melbourne (who) wish to prevent the larger mass coming to England.' In mid-July, 1863, he was making arrangements with the British Museum trustees to have the

Colonial Secretary send the Abel meteorite, now purchased by the museum, to Melbourne and 'consigned to some official person there, connected with the Colonial Office. It could then be handed over to the Melbourne Museum in the event of their fulfilling their engagement and giving us the larger mass in exchange for it.'[12] On 12 September, two days after Sir Henry Barkly's time as governor ended upon his completion of almost seven years in the service of Victoria, No. 2 was despatched from London on the *Benares*. The vessel would reach Port Phillip in early December but, to the consternation of many, the meteorite would vanish for 16 months.

10
Without a Violation of Good Faith

When the meteor shower
is coming
and space delivers
rocks the size of cows
what will your final pleasure be?
Chris Mansell 'Darling'

The Yarrabubba impact structure in the vast desert of Western Australia is a candidate for the oldest such formation in the world. So scoured by erosion that its visible remains constitute only a low central mound, called Barlangi Rock, and a posited crater boundary of 70 km diameter, it is estimated to be 2.2 billion years old. Zircon crystals at the site hold the key to its age. Each one contains a small amount of uranium, which decays into lead at a constant rate. Although incredibly tough, even zircon crystals will shatter under a large enough asteroid hit, and their lead is 'squeezed' out. This resets the tiny atomic clock to the time of the impact, and allows scientists to date the collision to within five million years.

A hump in the landscape, rather than a visible crater, sometimes indicates a cosmic strike. Impacts of sufficient force will generate pressures 100,000 times greater than atmospheric pressure at sea level, and cause an elastic rebound of Earth's crust. This can produce a central uplift portion called a 'rebound dome.' Yarrabubba and Woodleigh exhibit this form. Gravitational anomalies are also commonly detected at large impact sites. These occur when concentrations of variable density rock, relative to surrounds, appear — such rock will have a different gravitational signature to its neighbours. Detecting impact structures is a complex business.

Furthermore, it is hypothesised that impact events and mass biological extinctions have locked step in the last 250 million years. Certainly Chicxulub and the theorised 'impact winter' caused by its sunlight-blocking dust cloud aligns neatly with the catastrophic species decline at the end of the Cretaceous, 65 million years ago. Evidence suggests similarly coeval extinctions and large-body impacts in the late Eocene, at the conclusion of the Jurassic, and at the close of the Permian. Both Woodleigh and the Warburton craters coincide with a minor Gondwanan extinction further back, in the late Devonian, approximately 350 million years ago.

Most Australian impact sites are found deep in the outback. But two relatively young ones, the 35 million-year-old Flaxman and Crawford craters, adorn the folded geology of the Adelaide Rift Complex, just to the north-east of the South Australian capital. Exposed to the surface, there are tell-tale shocked quartz deposits secreted here, an indicator of metamorphism seen only in collisions of cosmic bodies. These two sites imply a low-angle trajectory for the impactors, which are possibly twinned pieces of the same body. To date, more than 190 asteroid impact structures and meteorite craters have been identified around the world, over 30 of which are found in Australia.[1]

September of 1863 found Mueller trying, yet again, to front the Board of Inquiry. 'I should feel grateful if the Board could be called together at an early day in order that my explanation may be received and the report be drawn up for final consideration of the Government,' he wrote to Evans on the 11th. Mueller was keen to update Maskelyne on progress, and:

> *I am also anxious to express to the Board my willingness to relinquish the position of representing on this occasion the interest of British Museum and to resign to Mr Bruce again the trust he imposed on me, even if it were only to show, that I have no personal interest whatsoever in the transmission of the Bruce Meteorite to London, but that I advocate this measure simply on universal claims of science.*[2]

Here the botanist, similar to FitzGibbon's motion at the Royal Society meeting the previous November, was taking care to note an absence of personal advantage in his connection to the meteorite. Like the town clerk, he emphasised that his goal was a positive outcome for science. FitzGibbon had only local scuttlebutt to consider when taking his defensive stance, but for Mueller the more official and unfavourable draft report was a weightier consideration. An argument with McCoy the previous evening also influenced his offer to withdraw.

At the time Mueller's residence was the elegant Gardens House at the Botanic Gardens, set in a silvan haven of shade and scent. When the two met there for dinner on 10 September the meteorite was discussed, and McCoy's simmering frustration with his colleague boiled over. Strong words were spoken. Mueller was aghast at the turn of events and next day wrote to McCoy in a conciliatory tone:

> *You were in such emotion when you left me last night, as to prevent me from apologising for the offence I unconsciously have given you. In my friendly remonstrance against continuing to discuss a topic, which evidently grew more and more painful to both of us, you gave certainly a wrong interpretation to my intentions, as nothing could have been further from me than to desire exhibiting a want of courtesy.*[3]

The Irishman had made his feelings about Mueller known in numerous letters to third parties over the meteorite issue, and doubtless also during this heated exchange, but Mueller was never intemperate in his remarks regarding McCoy:

> *Although I plead certainly innocent of such intention, I ask nevertheless most cordially your forgiveness for having wounded your feelings and to assure you that not many things in this world could cause me more grief than seeing the friendship towards me clouded of a Gentleman, and a fellow labourer in the empire of natural science, for whose discoveries and researches I entertain the profoundest regard, and whose acts of kindness have been towards me so manyfold, so genuine and extending over so many years.*[4]

The diffident Mueller did not want the quarrel to endanger the pair's friendship, and extended an olive branch:

> *I am quite willing to relinquish the position to whomsoever else chosen of representing the interests of the British Museum and of Mr Bruce in reference to the transit of the meteorite and to assign to them my trust again, even if it were only to show, that I have no personal interest in adhering to the arrangement into.*[5]

McCoy did not reciprocate Mueller's conciliatory tone, for later in the month Mueller wrote to Maskelyne that 'Professor McCoy evidently regrets having made the offer of the exchange, and it is painful for me to add, that by my remonstrations a friendship must become clouded, which has existed between us for many a year.' He also informed Maskelyne that upon receipt of Mueller's 11 September letter to Evans, the Chairman withdrew the draft report and at last called Mueller to appear. But once again the Fabian tactics of the board frustrated Mueller. What he called 'various circumstances' and 'a severe illness of the Honourable gentleman' (Evans) prevented the board interviewing him in September, and thus it was now Mueller who missed the monthly mail, and the chance to inform Maskelyne with 'the final view of the Board and the Government's opinion thereon.'[6]

Despite his success at the Royal Society meeting of 17 November 1862, the subsequent equivocations of the board and the tenor of its interim report meant Mueller still toyed with the thought of renouncing his agency of Bruce and the British Museum. As a Victorian public servant he was required to follow governmental instructions. He told Maskelyne that 'should the ministry express a disapproval of the arrangements entered into, I shall, as an Officer of the State be obliged to resign my trust again to Mr Bruce.' In that event, he would not fail to 'lay before Professor Owen and yourself a status of my proceedings and to afford to whomsoever Mr Bruce may appoint to act for him as well as to your future representative every information on our transactions.'

Events of October 1863 would change all that. On the 20th of the month Mueller finally appeared before the board of inquiry. Some surprising news greeted him:

> ... *That, having closed their labours in bringing up their report, the Gentlemen constituting the Board could receive no further*

> *information and considered the Board not further in existence and regarded for any of its further actions it necessary that it should be reappointed by Government. Therefore no legitimate meeting of the Board could have been held after the 20th October without this appointment.*[7]

Confounded and disappointed, Mueller notified Maskelyne five days later that 'my communication with the meteorite Board has been most unsatisfactory.' But this procedural impasse was not the only disheartening news Mueller garnered on the 20th. McCoy, he reported, 'never seriously proposed the exchange of the specimen' — something Mueller knew to be untrue — and Evans, like Royal Society member Charles Mayes the previous November, 'doubts the legal rights of ownership exercised by Mr Bruce.'

These were damaging broadsides but they served to bolster Mueller's wavering commitment. The previously humble supplicant was now galvanised. There would be no more talk of resigning from his intermediary role. Referring to some depositions the governor's office received from London, he told Maskelyne 'the communications received through the Home Government will probably strengthen me to act on your behalf,' and 'believe me, that the interests of the British Museum will not be neglected.'[8]

The tone of Mueller's correspondence with McCoy also changed, and the collegial tilt of his previous communications receded. In his letter of 28 October he dispensed with his usual address, 'my dear Professor,' in favour of 'dear sir,' and continued the text in a perfunctory manner. He was forwarding, on behalf of the new governor, Sir Charles Darling, the recently arrived Colonial Office despatch and enclosures 'bearing on the arrangements enlisted into for effecting the exchange of the Brucean and Abelian meteorites on behalf of the British Museum.' He continued:

> *I have further to observe, that his Excellency desires me to ascertain whether the authorities of the Melbourne Museum are prepared to give up the meteorite now in their possession on the condition of receiving that which is to be sent to his Excellency.*[9]

Brief and to-the-point. Mueller was cautious of McCoy after his humiliating interaction with the Board of Inquiry and McCoy's faithless behaviour on the meteorite issue. He had right to be. Unbeknown to Mueller, the Irishman was also working some sleight of hand with the No. 2 sample. The governor and Mueller were expecting Abel's specimen to be delivered to Melbourne, and were unaware that it had already arrived. The crown agents in London had addressed the consignment to 'the Trustees of the Museum Melbourne, care of his Excellency the Governor.' In Melbourne it was delivered to McCoy, who happily stored it away at the museum and told no-one about it.

McCoy also took his time responding to Mueller's request on behalf of Darling, waiting until 11 November to write and even then, still stalling:

> *I will not fail to have the information required to enable His Excellency to reply by next mail, to the Duke of Newcastle's letter prepared in ample time, but as the reply does not depend on myself alone it involves some grave considerations. I have been unable to reply directly.*[10]

A reply from McCoy would have needed to mention that he already had No. 2 in his possession, something he was not yet prepared to reveal. He had another plan afoot. On the next day McCoy wrote to James McCulloch, the energetic and capable businessman-politician who had succeeded John O'Shanassy as Chief Secretary of Victoria, and requested 'that you will have the goodness to cause an Authority to issue for the expenditure of the sum of £300 out of the vote for contingencies of the National Museum for the current year.'[11] This is the same amount previously referenced as the asking price of Abel's meteorite, and McCoy seemed to be preparing for its purchase from the British Museum, in preference to forwarding No. 1 in its place.

Charles Darling was a nephew of former New South Wales governor Ralph Darling, and, after his posting to Sydney in 1826 as an ensign in the 57th Regiment, was assistant private secretary to his uncle for several years. Without McCoy's input, the governor was briefed via a long letter from Mueller on 28 November. The botanist, apparently without irony, referred to it as 'a succinct statement in reference to the arrangements entered into for effecting the exchange of the Brucean and Abelean meteorites between the Melbourne and British Museums.' It cited no less than 18 enclosures, all letters pertaining to the detailed goings-on of recent years. One, from James

Bruce, must have used some fruity language, with the very proper Mueller withdrawing it in a postscript due to 'some uncontrolled expressions.'

There are no surprises here. Mueller wrote at length on the events surrounding the meteorite; the approach from Maskelyne, Bruce's original intentions ('*unrestricted* authority to effect the transmission'), the suggestion of division and the subsequent lapse of that offer, and his experience of the Board of Inquiry. Much as he cherished his adopted country and wanted to nurture its scientific and cultural growth, Mueller supposed the Bruce meteorite to be so extraordinary that his world view should prevail in this special circumstance:

> *But as I feel that I acted conscientiously throughout to the best of my judgement, I have no fear, that even if my own views were deemed wrong by the Government, my cosmopolitan tendency will be respected, a tendency to which on so exceptional occasion as this I cheerfully sacrificed that patriotic love, which I have imbibed for this country and the ardent desire of aiding in the rise of its institutions.*[12]

Mueller was ever mindful to do the right thing by his professional colleagues, even when such a singular specimen was in dispute:

> *Feeling nevertheless that some moral obligations were existing for honouring the claims of the Melbourne Museum and feeling an extreme regret that thus a possibility existed of so magnificent a Cabinet-piece being divided, I made an appeal to Prof. McCoy (vide letter no. 4) to waive his legal claims and interests in an instance like this in favour of the British Museum.*[13]

Events of December 1863 would draw Mueller's attention away from meteorites. In the fortnight before Christmas a heavy downpour over several days caused a prodigious flood in Melbourne and surrounds. Peak waters in the Yarra were the highest since 1835, in places up to 50 feet above average summer flows, driven by five inches of rainfall in three days and a southerly tempest that banked up the Port Phillip tide at the river's mouth. The *Age* reported on 19 December that 'European astronomers have predicted the calamity which has submerged so many fair fields around the Victorian capital, assigning as a main cause the Moon's proximity to the earth in the present December as a disturbing influence on the tidal waters.' The Moon *had* been at its closest to the Earth for the month on 12 December, but the limited effect of its perigee on tidal forces was not appreciated at the time. Robert Ellery and Georg Neumayer at the Observatory had earlier divined troubled weather ahead, and warned the population 'to look out for great disturbing influences in the elements in the middle of December.' In the midst of the torrential rain the Observatory recorded an air pressure of 29.003 inches of mercury on 15 December, a record low that stood for many years.[14]

After the flood of 1839 the Yarra was diverted in some sections from its original line, but now broke these levy banks and returned to sluice through its previous course. On its north side the Zoologic Gardens, where animals were housed in rough-and-ready enclosures at Richmond Paddock, were entirely submerged.[15] On the south bank, the bridge at the foot of the Botanic Gardens had its approaches washed away, leaving a stranded arch across the water. The gardens were inundated to a distance of 300 yards and the *Argus* reported them to 'have suffered greatly, the many beautiful shrubs and trees, which but recently were in full leaf and beauty, having disappeared before the flood.' Mueller had his work cut out with repairs and restoration.

* * *

In London, the redoubtable Maskelyne did not share Mueller's even-handed view of McCoy. When he responded to the botanist's communications of September and October, in a caustic letter written in early January 1864, he was seething. Although he opened with an insouciance inappropriate to the occasion, 'I have had two letters from you and I ought perhaps to have answered them,' he was careful to praise Mueller's conduct in the matter; 'In this I have the pleasant duty of having to thank you in the name of the British Museum for having maintained an attitude such as a German – or an English – gentleman must always wish to maintain, an attitude of honour and truthfulness.' The he proceeded to lambaste McCoy:

> *English Science feels not only friendly to, it is feeling and acting in a spirit of cordial alliance and unity with the Science of the Australian Colonies. I have not breathed a syllable on the disgraceful breach*

> of faith that one person – (and I cannot believe there are more in a community of Englishmen) – has been guilty of. I trust honour will reassert itself even in him.
>
> But I know that when I publish the facts and lay them before the Royal Society and the rest of the compact body of English men of Science, there will be such a revulsion of sentiment as must be felt in Melbourne. I must draw up such a statement and support it by documentary evidence, if this disgraceful business is not concluded in accordance with honour and good faith.[16]

Maskelyne closed with additional appreciation for Mueller's conduct, and one more back-hander for McCoy: 'one not acting with honour and generosity.' It was a forceful tirade, and much in keeping with Maskelyne's patronising view on the primacy of imperial institutions over their colonial models; his aforestated 'proper relative functions of mutual interest and support.' And is there an anti-Irish sentiment between the lines of his letter? To praise the noble attributes of honour and truthfulness in English and German gentleman, while pillorying one Irish-born, can lead the reader to an obvious conclusion.

Maskelyne's adversarial stance unsettled Mueller, and in his return letter of 25 March 1864, he pleaded for restraint. Although McCoy had irked and offended him, he was not prepared to see his erstwhile friend's reputation besmirched before so august a body as the Royal Society of London. 'Meanwhile, I would beg of you for the honour of Victorian Science to refrain from drawing attention of the Royal Society to the line of action,' he wrote, 'which here was adopted when a fairly and sincerely made promise was to be redeemed.' Mueller's position would try the wisdom of Solomon and the patience of Job, but Maskelyne's support fortified him. 'It is a great consolation to me to be conscious of having done throughout what I regarded fair and measured and above all to be assured of your recognising my action as conscientious.'[17]

Still unaware, along with Governor Darling, of No. 2's whereabouts, Mueller was uncertain whether Maskelyne and Owen 'may not have preferred to retain the Brucean meteorite until further communications from here' (a mistake; he meant to write 'Abelian meteorite.' This mix up of the samples was repeated in a subsequent letter from Mueller, but with the incorrect reference struck out this time. Mueller penned a very large number of letters). Darling, armed with Mueller's extensive summary and enclosures, sought more information from McCoy. He then, as reported by Mueller to Maskelyne in late April, 'transmitted all documents connected with these transactions previously to the Ministerial Cabinet, disapproving of the reasons on which the meteorite was retained, as set forth by the Meteorite Board.'[18]

Darling's gathering of relevant documents led to the conclusion he presented to McCulloch on 12 April. He expressed sympathy for the board's 'anxiety that a valuable object of science should become the property of the Victorian Museum.' However, he 'altogether dissented from the reasoning upon which it founds a claim on behalf of that Museum to one half of the Cranbourne Meteorite.' He cited James Bruce's first letter to the *Argus* in December 1862 as the fulcrum for his decision:

> *The papers herewith seem to me conclusive as to the intention of the persons in whom the property in that meteorite resides that it should be transferred to the British Museum and if there is one document which more than any other strengthens this view it appears to me to be that upon which the Meteorite Board relies as containing the promise to allow half to be retained by the Victorian Museum, viz Mr Bruce's letter to the Editors of the Argus of the third of December 1862.*[19]

Mueller reported to Maskelyne that upon finally reading the board's report, sent to him by Darling, he corrected 'a series of errors contained in this document, which could have been avoided, had I been allowed to tender my evidence.' Still smarting at his treatment, he recommended 'that Professor McCoy should show cause why the pledge made by himself should not be redeemed.' Furthermore, 'that unless satisfactory reasons by him were given for the withholding of the meteorite, the Government should arrange through use of the Engineer Offices of this state the earliest possible transmission and charge the transit expenses to the Home Government.'[20]

In addition, the Lanarkshireman McCulloch came down on the side of his fellow Scot, Bruce, but without overt parochial leanings. The Chief Secretary had already duelled with John Macadam on the floor of the

Legislative Assembly in May, in response to the good doctor's provocation that 'it would be a disgrace to the colony to allow the meteorite to be removed.' Macadam was still promoting William Crooke's watered-down amendment, moved at the Royal Society meeting back in November 1862, and trying his best to 'secure possession of the Bruce meteorite for the colony of Victoria.' But McCulloch asserted the government's position – that it had no power to interfere with the meteorite's owner. The member for Normanby, George Levey, put up a smokescreen regarding crown rights to ownership, and equating flotsam on shore with articles 'which found their way to the earth from the heavens.' George Evans weighed in with a recommendation, rather late in the day, to source the opinion of the crown law officers.[21]

Also around this time a back-channel attempt was made, apparently on behalf of the government, to persuade Bruce to forego his claim. But even an entreaty from Alexander Patterson, Cranbourne area notable and leaseholder of the St Germains run by Cardinia Creek, could not move him. The two were fellow members of the Cranbourne Road Board before Bruce's departure, and Patterson would go on to lengthy service as shire councillor and president. In February 1865 Patterson told an audience, gathered in Clyde to hear his lecture on astronomy, that the Victorian government had charged him the previous year to contact Bruce – now in England – and request the retention of No. 1. An inducement of 1,000 pounds supposedly accompanied the appeal, but the owner refused, saying 'money would not buy it.'[22]

Such an offer appears inconsistent with the actions of the premier. For McCulloch the situation was simply cut-and-dried, as he replied to Darling in June:

> *Afterwards he (Bruce) refused to permit the meteorite to be cut and requested Dr Mueller to dispose if as first intended viz to be presented to the British Museum. I do not see on what grounds even an attempt can be made to prevent the wishes of the owner being carried out no doubt it would have been desirable to have retained the meteorite in the Colony but that cannot now be done without a violation of good faith. The meteorite can therefore in my opinion be only treated as the property of the British Museum subject of course to any outlay there may have been for carriage etc.*[23]

With McCulloch willing to gainsay the recommendation of the previous government's committee, and Darling weighing in on the matter in Bruce's favour, the push to consign No. 1 to Britain was gaining considerable momentum. In July 1864 Darling wrote to Edward (later Viscount) Cardwell, Secretary of State for the Colonies and an old Colonial Office hand, to report 'I am in a position to announce, that the meteorite now on this Colony is regarded by the Colonial Government, as the property of the British Museum.' But where was No. 2? Darling laid out what had been arranged between Cardwell's predecessor, the Duke of Newcastle, and the British Museum trustees for the transfer of Abel's meteorite. It 'would be forwarded to this Colony addressed to my care, which mass I am instructed to retain, and await further directions.' That object, however, 'had not reached me.'[24]

In October 1864 the Colonial Office quizzed the crown agents, Messrs Julyan and Sargeaunt of Spring Gardens, about No. 2, quoting Darling's inquiry of 20 July, and asking why 'the meteorite belonging to the Trustees of the British Museum which you were instructed to send out to the Colony, had not reached its destination. Mr Cardwell would be glad therefore to be informed of the date at which the meteorite was shipped by you and whether you can furnish any explanation of the delay attending its arrival.' William Sargeaunt responded the next day with the date of despatch of the secondary meteorite, 12 September 1863, but could offer no explanation regarding its whereabouts.

As it happened, Frederick Mueller and Governor Darling were not made aware of No. 2's location at the museum's University grounds until December 1864, more than a year after its arrival. Mueller immediately urged action 'to respect the claims of the British Museum' and Darling subsequently issued an order to the Government Storekeeper for No. 1 to be removed and transported to London.[25] In the middle of January 1865 the *Argus* reported on the National Museum's visitor numbers (McCoy's mining models were a popular attraction) and advised 'the great meteorite ... may be shortly sent to England. Colonists desiring to see this most remarkable specimen can probably have only a short time for doing so.'[26]

The newspaper was right. But McCoy still seemed caught unawares when 'two contractors giving the names of Hurley and Carrol' arrived at the museum on the afternoon of 21 January 1865, to collect No. 1. If he appealed to his fellow Irishmen to stay their hand he was not heeded. But 'they promised not to ship it until an order was brought to me from the Chief Secretary authorising me to give up the specimen, and stating which of the 2 meteorites was to be removed,' he wrote hurriedly to the Chief of Detective Police that day.[27]

McCoy was combative to the last. Even now he attempted to sow confusion as to which of the two meteorites was to be handed over, and shamelessly referenced the now-revealed No. 2 as possibly the removalists' intended target. As well as requesting police assistance he wrote to McCulloch suggesting that 'as the Board appointed by the late government recommended that this specimen (No. 1) should not be parted with I thought it probable that it was the small one the property of the British Museum which should have been shipped.'[28]

McCoy wrote to the Keeper of Stores two days later, on receipt of a letter from that office asking permission to remove No. 1, cleverly written on the same day as Hurley and Carrol presented at the museum quoting their verbal order. McCoy complained that this letter 'would have been returned by me with a note that I could not deliver the specimen which was in my charge without a letter from the Hon. The Chief Secretary in whose Department the Museum is, and who is my superior officer in such matters authorising me to deliver it, and specifying which of the meteorites should be shipped.' McCoy well knew which specimen was to be transported but he kept up the pretence of ignorance, and introduced a red herring about No. 2 being shipped back before demanding No. 1's return:

> *The second meteorite alluded to is now the property of the British Museum sent out to Sir Charles Darling with advices saying it was to be exchanged for the large one, or if there was any objection to that it was to be shipped back again.*
>
> *… I hereby require you to return it to the office from whence you took it until I have been provided with a formal authority from the Chief Secretary to act in the matter. The second meteorite being the property of the British Museum might be dealt with otherwise.*[29]

McCoy, who had so often cut his coat to suit his cloth in curating the museum, was now trailing the garment as well. But all of the Irishman's confected outrage came to nought, and there would be no letter from Chief Secretary McCulloch advising of a favourable outcome. The *Red Rover*, a clipper of 1,100 tons which had undergone a partial refit since arrival in Melbourne, was docked and awaiting placement of its meteoritic cargo. On Saturday 11 February with No. 1 loaded snugly in its hold among several thousand bales of wool and 3,000 ounces of gold, it raised anchor and set sail for London.

Mueller delightedly informed Maskelyne of developments; 'It is, dear Professor, with extreme gratification to me that I am able to inform you of the instructed shipment of the Brucean Meteorite by the 'Red Rover' which ship is to leave our port in the early part of February,' he wrote on 25 January. 'If no disaster occurs to the ship on the voyage I hope you will have your treasure in May next.' He had reason to be ebullient after his multi-year struggle to secure No. 1 for the British Museum, and Maskelyne's supply of a personal portrait chuffed the sentimental scientist; 'I feel very proud in the possession of your miniature photograph and shall always value it as a token of your friendship.' Perhaps by way of reciprocation, in a touching aside Mueller offered his botanical works to Maskelyne's wife, the astronomer and botanist Thereza Llewelyn. In August 1858, while on her honeymoon, she had observed Donati's comet before the announcement of its discovery by the Italian astronomer.[30] 'I have published 8 volumes on Australian plants' wrote Mueller, in a cramped postscript in the margins of his letter. 'If any are desired by your Lady I will be happy to send them.'[31]

Maskelyne, Owen, and the British Museum had been well served by their intermediary. Despite his commitment wavering after the frustrations of the board of inquiry, Mueller had continued to pursue their shared objective in the face of overt and determined opposition. The meteorite was 'treasure' indeed, but would no longer crown the collection of Victoria's National Museum in its country of origin. Sadly for Victorians, had anyone but Ferdinand Mueller been charged with the British Museum's mission, there may well have been a different outcome.

11
The Honour and the Gain

Looking up, the galaxy senses itself, Gemini points out a neck cross, Lynx bones wind up down along the Great Dividing Range.
Peter Minter 'Knitcap Sutras'

One hundred years after the Cranbourne affair, another celebrated meteorite visited Victoria. This time it was a sighted fall, and yielded specimens of rare composition that drew attention from around the world. The town of Murchison, founded in 1840 as the Goulburn River District Aboriginal Protectorate, lies on a bend in the Goulburn River to the north of the state, 170 kilometres from Melbourne. Explorer Thomas Mitchell passed by to its south during the return leg of his 'Australia Felix' expedition in 1836, when he temporarily entitled the Goulburn with the Dayngwurrung name 'Bayungun.'

The year 1969 was already replete with space age imagery, and it was mere months after humans first walked on the Moon when, in the late morning of Sunday, September 28th, the sky above Murchison was breached with a bright flash and tremendous explosive reports. In nearby Kialla West a vivid orange ball with silvery rim and dull orange tail was seen tracking in a north-westerly direction. The interruption startled dairy farmers and churchgoers alike, compared by some to an express train and to others, a jet aircraft. High up, three pieces were observed to separate from the bolide, and they beat a percussive tattoo upon shattering the sound barrier. Livestock cowered in the fields and the clear sky was smeared with a trail of blue smoke.[1]

Fragments then showered over Murchison and surrounding areas, some landing in a farmer's yard, others on a fairway of the local golf course, and more in a hay field. One piece crashed through the roof of a haystack, coming to rest on the silage below, perhaps the softest terrestrial impact of any astronomical body. The resulting strewn field covered an impressive eleven kilometres by three kilometres.

It was a thrilling event for the town, fortunately devoid of injuries, and residents were swept up in the excitement. The University of Melbourne was involved early, with the head of the Geology Department, John Lovering, organised groups to help in recovery. The search for fragments took on the air of a harvest fair treasure hunt, with young and old vaulting fences and scouring the countryside in search of space rocks. No stone was left unturned; resolute students even examined sludge from the hosed-down apron of a cow cocky's milking shed. The bulk of the 100 kg of specimens recovered was assembled in short order. Murchison is especially valuable because the speed of its fragments' collection means little, if any, terrestrial contamination was incurred.

The meteorite captured immediate global attention with the boisterous manner of its arrival and the exquisite timing of its entry – the universe 'pushing in' just as humankind first stepped out on another astronomical body. It's ironic; outer space delivering rocks to Earth by the dozen, immediately after humankind had spent enormous sums going to the Moon to retrieve space rocks. It continued a theme begun earlier in the year. On February 8th in the northern Mexican state of Chihuahua, the occupants of the village of Pueblito de Allende were treated to a brilliant bolide that rained thousands of fragments – carbonaceous chondrites – over a 400-square-kilometre strewn field.

After the initial burst of publicity, Murchison sustained attention with the revelation of its intriguing mineral structure. As well as the small mineral-grain chondrules within its matrix, it contains blue-coloured hobonite crystals, rich in calcium and aluminium, which were formed into rocks from molten droplets as the early Solar System's spinning disc slowly cooled. Trapped inside these are isotopes of helium and neon, split from minerals by the young Sun's intense radiation.

Like Allende, Murchison is classified a carbonaceous chondrite, one of the most primitive of meteorites. Carbon-rich, it formed close to, or during, the birth of the Sun and so maintains the chemistry of the earliest Solar System, even holding attributes of the pre-solar molecular cloud.[2] Comprising a pudding of hydrated clay minerals, it has a relatively high water content; up to 10%.[3] Solvent odours detected in the samples first suggested the presence of organic molecules. 'It smelt like metho,' suggested one local upon his encounter with a fragment, drawing a comparison to methylated spirits. Subsequent examination by mass spectrometry revealed thousands of unique molecular compositions, and possibly millions of distinct organic compounds, in the meteorite.[4] Of particular interest was the occurrence of amino acids, the building blocks of proteins, in Murchison specimens. Some included right-handed molecular symmetry, a variety never encountered on Earth, and accelerated discussion on the possible meteoric 'seeding' of our planet with these essential ingredients of life.

The Apollo 11 astronauts collected 21 kilograms of rock and soil during their vaunted mission in July 1969, and the six Apollo lunar missions accounted for a total over 380 kilograms. Murchison has been arguably the greater contributor to the scientific canon. It continues to fascinate scientists, fifty years after it blasted through our atmosphere.

The sailing of the *Red Rover* with its precious cargo represented a telling loss to Frederick McCoy and his like-minded colleagues in the Royal Society of Victoria. Cranbourne No. 1 was a glittering prize, and it had slipped through their fingers. The palaeontologist's protests fell mostly on deaf ears, although the *Argus*'s town and country weekly, the *Australasian*, gave him a sympathetic hearing in late January 1865:

> *We are informed that, on Saturday afternoon, the large meteorite which so long formed the most remarkable and famous colonial specimen in the public Museum was removed, without acquainting the director, in whose charge it was, by three persons purporting to act on verbal orders of the Government storekeeper, to ship it on board the Red Rover, to sail on Monday morning. A formal protest was made against the removal without a written authority from the Chief Secretary.*[5]

Little else was done to publicise or protest the forfeiture: no groundswell of objection, no petitions nor further agitation. The anniversary address of Royal Society President the Reverend John Bleasdale, delivered on 4 May 1865, did not even reference the meteorite, and the society's transactions are silent on the subject after this time. The public's attention was elsewhere, distracted in part by newspaper accounts of the murderous exploits and eventual deaths of bushrangers Dan Morgan, Johnny Gilbert and Ben Hall early in the year. Cranbourne slipped from view.

The meteorite arrived unscathed in London and was installed at the British Museum in the first room of the Mineral Gallery, uncased and fixed to a turntable. Soon after, when fragments were crumbling off and it was found to be rusting at a concerning rate, a coating of shellac varnish was applied and a glass case arranged for its protection. Trays of regularly renewed caustic lime reduced the damaging oxidation process.[6] When the museum's 1865 accounts were laid before Parliament the following year, the *Times* listed its important acquisitions, including more than 30,000 natural history specimens for Professor Owen's department. Cranbourne No. 1 was described as 'the most remarkable addition that has ever been made to the collection of meteorites.' Even so, attendances to the general collections were down by 14 percent on the previous year.[7]

The dispute over the Cranbourne meteorite was a drawn-out affair; disconcerting, even painful, for some participants and a strain upon professional and personal relationships within Victoria's scientific community. The unfavourable outcome does not diminish the concerted efforts by McCoy and his colleagues to retain a unique natural history object within its country of origin. When viewed from the contemporary perspective one may not appreciate the boldness of the attempt, nor McCoy's audacious guerrilla tactics, during a time when many colonists accepted the pre-eminence of British home institutions. That some who professed local sympathies still wished the meteorite transferred offshore can be disappointing to the modern reader, but perhaps it is naive to expect far-sightedness of these actors from an earlier time. Britain's 'imperial century' was at its halfway mark; in addition to an unchallenged navy and dominant world trade position, it exerted extensive control over its colonies. The violent flaring of rebellion in India, and its harsh suppression, was a recent memory. Colonial possessions, outliers on the imperial board, supplied the core with trappings and treasure. Australian wool fed the mills of Britain's midlands, surpassing German imports there by the early 1840s and delivering more than half of British imports by that decade's end. Throughout the 1850s and 1860s, gold exceeded even wool as

Australia's premier export earner, with over 25 million ounces shipped since 1851. The output of the revolution in British industry was based in part on Australian toil and produce.

Despite this economic interdependence, the establishment of scientific foundations in Melbourne signalled the beginning of separation between metropole and satellite. But the early rumblings of autonomy from the scientists of Victoria did not vibrate sufficiently to move Nevil Maskelyne. His references to Calcutta and Madras, and their contributions to the British Museum's enrichment, is a reminder that that the 'proper relative functions of mutual interest and support' was a one-way street that led into Bloomsbury, not away from it. The haughty dismissal of the Melbourne gentlemen who proposed division of No. 1 is provocative. Why denigrate the skill and achievement of local scientists? After all, they were transplanted products. Colonial Australia's civic and academic culture was modelled squarely on metropolitan institutions. McCoy, Selwyn and Macadam had received their educations at Oxford and Cambridge; to consider them inferior was to relate distance with quality, an inappropriate association.

British imperialism spanned scientific disciplines as well as political and cultural contexts.[8] William and Joseph Hooker promoted the traffic of botanical specimens from all points of the compass to the gardens at Kew. Charles Darwin requested biological samples of interest from contacts in countries far and wide, and expected acquiescence. As early as 1830 the Royal Society and other learned institutions in London had instructed one of their fellows travelling to Van Diemen's Land, Matthew Friend, to promote the sending home of colonial specimens. Friend, with expertise in astronomy and natural history, told a meeting of the Van Diemen's Land Society that 'You are in a country whence much scientific wealth may be obtained. You are the possessors of a rich mine not yet assayed. Such of the ore as you find yourselves unable to reduce, we request you to send to us.' Like Maskelyne 30 years later, he spoke of 'reciprocal benefit' but was vague on tangible benefits to Hobart and its hinterland. No matter, there was honour in the suggested course of action; 'It is probable that our greater experience, and our more numerous and better fitted appliances will enable us materially to facilitate your labours.

The honour and the gain we will divide betwixt us.'

The honour and the gain indeed. Would they be *equally* divided, however? Maskelyne and Owen were instructive in directing colonial governments to be on the lookout for meteorites with which to enhance their museum. Imperial decree held sway over colonial assets, and the source would not be surpassed by its satellites. In this manner nine meteorites collected by the British Museum between 1859 and 1870, including the Parnallee chondrite and a rare Martian specimen, from Shergotty, were sent to London via the office of the Secretary of State for India.[9] These interactions were cloaked in the language of 'exchange,' but despite Maskelyne's view on the colonials' position, that 'they gain more than they lose, and so do we,' some fossil casts and duplicate aerolites sent from London could not replace a singular local specimen, its connection to place, and its position within a wider cultural circumstance. Maskelyne's ad hominem criticism of McCoy – no docile colonist – in his January 1864 letter to Mueller demonstrates his tin ear regarding local concerns. The cultural hegemony would continue, and habits of years would not change quickly.

Even as ardent an internationalist as Mueller was expected to comply with the status quo, when he was the obvious candidate for authorship of the Australian flora volumes, the *Flora Australiensis*. It was a recurring theme, of sorts. Like Alfred Selwyn, a local geologist named C.A. Zachariae had argued against Roderick Murchison, in this case in the *Bendigo Advertiser*, for his conflation of Ural and Australian gold. As part of his response Murchison declared 'judging from the language he uses, I apprehend that Mr Zachariae is a foreigner, who cannot adequately convey to English readers a correct conception of his own views'[10] Not without cause had Benjamin Disraeli earlier referred to Murchison as a 'stiff geological prig.' If Murchison could sneer at the alien nature of Zachariae, then a German was unlikely to find favour over an Englishman in the matter of important British scientific publications. Perfidious Albion prevailed. Mueller did as he was told.

The presumption of entitlement was not confined to institutional halls of power. McKay's and Bruce's assumption that No. 1 was theirs to barter was neatly aligned with European 19th century thinking. Possession meant ownership, and indigenous claims – if they could even be expressed in a manner recognisable to settlers – were secondary at best, and non-existent for the most part. James Bruce's actions were an inadvertent imitation of his namesake Thomas Bruce, the 7th Earl of Elgin. The controversial nobleman had, earlier in the century, arranged to strip the Parthenon in Athens of many of its Phidian sculptures, for eventual sale to the British government. And although they were later advocates for the retention of Bruce's meteorite, FitzGibbon's and Selwyn's opportunistic sampling of No. 1 while still in the ground may have presaged Schliemann at Troy, smuggling Priam's treasure out of Turkey for placement in Berlin's Pergamon Museum. Cultural appropriation could be explained away by the argument, asserted or implied, that yours was the worthier of two disputing cultures.

Elgin and Schliemann were only two examples of barefaced looting, where the overbearing assumptions of superiority of 'old Europe' protagonists swept aside nascent concepts of national identity and attachment to cultural icons in countries at Europe's edge. Cranbourne, on the other hand, was an episode of intra-empire struggle, where the satellite was pitted against the primary. It was a very British affair, Mueller's and Neumayer's involvement notwithstanding, and its cast was a broad range of actors with competing priorities.

Initially the Cranbourne specimens were not treated as cultural treasure, but as nuisances or potential mementoes. And so Lineham bartered his away, and FitzGibbon, Neumayer, and Selwyn took their souvenirs. Those who subsequently lobbied for No. 1's retention, and bemoaned its removal to London, viewed the loss from a local perspective, in that the colony was being exploited for its scientific artefacts. The cost was probably greater to local science than that to the burgeoning cultural identity of the colony. When the cultural aspect was explored, it was from the European viewpoint. That is, the meteorite was lost to the colony and its settler inhabitants, not removed from any possible Aboriginal sense of connection. The specimens empowered an emotional reaction from the Boonwurrung, judging by the two surviving references to their interactions: one, a joyous occasion where people capered around No. 1 and further engaged with it by striking it with their axes, the other an anguished episode with Aboriginal women wailing in lamentation over the removal of No. 2. One can conclude that such a palpable connection would also mean the conferral upon the meteorite of an identifying label; the Boonwurrung probably gave it a name.

With the indigenous perspective disregarded, only European protagonists, colonial and home-shored, remained. James Bruce called upon the Royal Society members who endorsed the retention of No. 1 to 'take a more cosmopolitan view of the matter, and lend their aid instead of throwing obstacles in the way.' As well he might; the voyager who was about to return to Britain would soon sever his local ties and lease out his properties, and it served his putatively righteous position to encourage a broader outlook. Mueller held a similar view, but without the luxury of professional removal from the main stakeholders enjoyed by Bruce. Active as both benefactor and recipient in a global network of scientific patronage, the German was firmly, if not always comfortably, in the removalists' camp from the moment he received Milligan's letter requesting assistance on behalf of the British Museum.

For one who set such store by integrity, Mueller's treatment at the hands of the Board of Inquiry perplexed him. He could be brought low by such actions. Only months after achieving the dispatch of Cranbourne No. 1 to London, when he should have been basking in the success of that arduous enterprise, his emotional scaffolding seemed unsteady. He wrote to William Hooker, describing 'my often difficult and dark path of life,' and spelling out grievances. 'I have still often to sustain injustice, and want of generosity, and both give me always great pain, discouraging me often for a while to do anything.'[11]

A more self-absorbed individual may have retreated after such let-downs, but Mueller invariably sought solace in his work. As a German making his home and vocation in a British colony, a dutiful brother concerned and responsible for his sisters' welfare, his personal path in life seemed a lonely one. Focused as he was on his professional career, Mueller still managed to be twice betrothed, although he never reached the altar. A third suit appeared well advanced before it too was discontinued. A colleague was dismayed when he received a letter penned by Mueller on Christmas day in 1865. But the writer may well have celebrated the holiday in the German tradition, on the previous day, most likely absent his now-married sisters and their families living in country Victoria and South Australia, and was using the festive season to catch up on his correspondence.

Mueller's regular obeisance grates, and one can be frustrated by his fawning to more senior figures who lacked his lofty intellect; however, his motivations were complex. His personal code mandated fairness and honour as worthy guardrails, but he formed some behaviours according to their perception by others, both peers and superiors. In one exchange with Maskelyne he pointed out the consolation he took from doing what he regarded as 'fair and measured' and 'above all to be assured of your recognising my action as conscientious.' Mueller certainly felt the eyes of others upon him, and his self-esteem could rise and fall with those opinions.

After his return from the Gregory expedition in 1857, and his appointment as director of Melbourne's Botanic Gardens, he set about establishing the National Herbarium of Victoria, contributing much of his own collection to its foundation. His bust stands outside that building today, a brass and copper likeness capturing arched eyebrows and Van Dyke beard to imposing effect. The knightly insignia are on show (Mueller received his German ennoblement in the late 1860s, and his KCMG in 1879) and other decorations clutter his lapels. The accompanying plaque on the granite plinth refers to Mueller as one of the fathers of the Garden State, as indeed he is.

* * *

The presumed supremacy of the metropole regarding placement of pre-eminent specimens such as Cranbourne, with which Mueller concurred, made Frederick McCoy chafe and bridle. Nevertheless, McCoy held home institutions in high regard; the Geological Survey of Great Britain was 'the finest in the world.' He just happened to think that colonial exertions were of comparable value. Describing Selwyn's efforts as government geologist he referred to the 'first-class excellence of the work' and, of the geologist's map and sections, 'they are on the same plan as those of the British Survey, and are quite worthy of being placed beside them.'[12]

When contrasted with Mueller, the Irishman presents the more interesting figure; something of a schemer, a stirrer, and one artful in the manipulation of higher-ups. A pioneer in palaeontology and biostratigraphy, he could classify huge collections, and doggedly advance any undertaking for which he was responsible. He maintained a rearguard action over many decades against the forces occasionally marshalled to deny him funds for his museum's upkeep and expansion; these included the University building committee, the Chancellor and Vice-Chancellor, the Registrar and professorial board, and the University council.

McCoy's chicanery ensured No. 2's location was kept secret upon its return to Melbourne. Yet again, he ran his own race, and normal protocols were discarded in favour of a personal agenda. His brazen heist was likely to keep alive the dialog for retention of No. 1. To have already received the quid pro quo represented by No. 2 would have weakened the argument for retention of the primary specimen. Its concealment from Darling and Mueller was the safest course, a scorched-earth act of omission that could later be explained away. Australian artefacts of naturally history, mineralogy, and geology were reasonably plentiful and so their supply to, and exchange with, international institutions was common practice. Astronomical samples were much less abundant, and a meteorite as novel as Cranbourne was rarer still. This was a special case. Even so, it took resolve to defy Maskelyne and Owen and their hidebound culture of entitlement. They assumed possession by imperial fiat, and McCoy was having none of it. His shortcomings – a sometimes casual observance of expected behaviour, for instance – lend a picaresque slant to his character, and reflect in part the general prejudice of Irish characteristics prevalent at the time. He certainly got under Maskelyne's skin.

The tragic deaths of three of his infant children marred McCoy's personal life. His surviving son, Frederick Henry, delivered eight grandchildren but pre-deceased McCoy by 13 years. The only daughter to reach maturity, Emily Mary, lived in the parental home all her life and also died before her father.

McCoy held sway over forty years as State Palaeontologist, Professor of Natural Sciences in the University of Melbourne, and Director of the

National Museum of Victoria. Despite the obligations of his professorial and curatorial duties he produced two major works during his time in the colony: the *Prodromus of the Palaeontology of Victoria* and *Prodromus of the Zoology of Victoria*. He was a competent teacher, but to the chagrin of his students he favoured classroom learning over field-based instruction, a reflection of his theoretician bent. It was a characteristic that became a pejorative as his tenure matured and younger scientists of more practical focus entered academia.

Being so long in the public eye, and not averse to expressing strong opinions, McCoy drew his share of criticism across the years. A protracted dispute was conducted with Rev. William Clarke, another Sedgwick protégée, over the age of some New South Wales coal fields. McCoy relied on evidence within samples supplied by Clarke to Cambridge when he was working there, and a European geological interpretation; Clarke had a stronger focus on field-work and persistent compilation of local stratigraphic data. The clergyman's position prevailed. In 1856 McCoy had also sided with Sir Roderick Murchison over the viability of deep-seam gold deposits in Australia, a position strongly, and correctly, contested by Alfred Selwyn and borne out by subsequent field evidence. Even with these occasional reverses, McCoy's renown steadily accumulated over his almost fifty years among the scientific leadership of Victoria. He was knighted in 1891.

Mueller and McCoy. Both were opposed to Darwinian theory upon its publication, and remained so. Both resorted to use of their own private funds, when government resources were unavailable, to supplement the collections they oversaw. Both were adherents to the Acclimatisation Society's objective of introducing and distributing foreign species to enhance local wildlife and livestock. They eagerly promoted the importation of exotics which, as McCoy wrote, were 'likely to thrive and do well when turned out in the colony.' That two such eminent scientists did not anticipate the harmful threat of this pursuit is considered a blemish by the modern viewer, replete as we are with ecological and environmental comprehension. It is understandable in a 19th century context, however; a time when extended periods of observation had not yet elapsed, nor contributed to the judgement of deleterious effects, and the nostalgic pining of colonists for the sights and sounds of 'home' was particularly strong.

Mueller and McCoy. After decades of service to the colony of Victoria they would be showered with honours locally and abroad, and achieve professional status, even fame, most likely unavailable had they remained in Europe. Their tussle over the meteorite appeared to not impact their professional and personal relationships, to the betterment of Victorian, and Australian, science, and they always had more in common than that which divided them.

In the end the duty-bound concept of honour settled their dispute. This complex code – tied to responsibility, manners, etiquette, and class – permeated thought and action of 19th century gentlemen. Its measure could be self-evaluated, such as Mueller's efforts to be fair and conscientious. Or others appraised it, as when Maskelyne saw the characteristic manifested in Barkly but declared it absent from McCoy's make-up. Chief Secretary McCulloch identified the attribute by defining its breach, i.e., 'a violation of good faith,' when describing attempts to pry No. 1 from the grasp of its owner. James Bruce's claim to the meteorite was proper in his view, and accompanying it was the right of disposal. Edmund FitzGibbon, when presented by Mueller with irrefutable evidence of Bruce's ownership and intentions during the Royal Society meeting of November 1863, ceded the field.

* * *

How should the words and actions of this cast of players be interpreted? There is McCoy, whose willingness to fib, delay, and obfuscate could have led to his professional undoing. And Mueller, whose sense of honour held him so firmly to a path that he was willing to risk a personal friendship. Barkly, an eager amateur attempting to umpire the vigorous feud of professional scientists. FitzGibbon, perhaps wanting to address his earlier acts of omission by renewed exertion to secure No. 1. The gruff Selwyn, the shrewd Smyth, George Foord with his empirical exertions and the popular Richard Daintree. And Georg Neumayer, ever the youthful beau sabreur; one moment riding with Burke and Wills to the limits of settlement, the next working his magnetic magic over the main Cranbourne specimen. It may be said that each simply did his best when faced with an unusual and challenging dilemma.

As it transpired, one's best efforts in this unique scenario varied widely between protagonists. In hindsight two events stand out as turning points: FitzGibbon declining to purchase when McKay and Lineham proffered the meteorites, and McCoy's failure to meet Bruce's request for a timely response to his suggested division of No. 1. The first, had it gone the other way, would have resulted in the samples, in their entirety, ensconced in the National Museum of Victoria. The town clerk later wrote of his regret – rueing the opportunity his decision gave Augustus Abel in 'getting himself glorified' by lending his name to the smaller block, and expressing outright disgust that James Bruce sent his meteorite out of the colony. The second's alternative scenario would have kept half of No. 1, and No. 2 entire, on Victorian soil.

And what of the much-stated benefit of home country placement touted during the protracted to-and-fro, particularly that this would be advantageous to more men of science? It was slow to manifest. In fact, little detailed analysis was performed on No. 1 in the decades immediately following its arrival in London. The first European publications were Haidinger's reports of 1861/62 to the Viennese Royal Academy of Sciences, written in German and based on Barkly's gifted piece and private correspondence from Neumayer. Haidinger's findings signified the start of research on Australian meteorites.

Also in 1861, Hamburg physician Karl Zimmermann reported in a letter to the German-language *Mineralogy Yearbook* on Abel's initial experiment on No. 2. The owner identified a new metal he called meteorine; 'mica-like lamellae, which spread in the mass in very thin, shiny rhombic leaves of almost silver-white color.' The industrious George Foord's observations of specific gravity and crystalline texture supplied to Smyth – 'Widmanstethian figures were beautifully developed' – were published in 1869.[13] But papers from Munich mineralogist Karl Haushofer, drawing on Barkly's sample, and French chemist Pierre Berthelot, using samples supplied by Maskelyne, continued the predominantly Continent-based appraisals of the early years.

And so No. 1 lay idle and rusting in its new home, more hollow trophy than hallowed treasure, and the Victorian Board of Inquiry's warning that 'in the British Museum it would be little more than an object of idle curiosity from its great size' seems a perceptive one. It was not until 1882, seventeen years after its arrival, that the museum undertook a detailed study. Walter Flight was a Winchester-born chemist and mineralogist, appointed as assistant in the mineralogical department under Maskelyne in 1867, and an emerging authority on occluded gases in meteorites. His article covering the Cranbourne sample was read before the Royal Society of London in February 1882:

> *The Bruce meteorite consists entirely of metallic minerals, and contains no rocky matter whatever. The iron contains no combined carbon, but from 7 to 9 per cent, of nickel, some cobalt, a little silicium, and copper; and, distributed through its mass, rather less than 1 per cent, of bright, apparently square prisms of a phosphide.*[14]

Flight identified the kamacite and taenite alloys in his specimen, in a later study using the delightful alternative names 'beam iron' and 'fillet iron' respectively. He believed his recognition of the latter, Abel's meteorine, to be the first such definite identification, despite the previous work of chemist Carl Von Reichenbach in this field, and summarily proposed endowing his own choice of name on it: Edmondsonite, after the headmaster of his Hampshire alma mater.

In Victoria, scientific publications made a similarly unhurried appearance. The first extensive paper on the Cranbourne specimens, by Richard Walcott of the National Museum, was not published until 1915. Even when four additional irons in the Cranbourne strewn field were revealed in 1923, and another in 1928, they were not reported on in detail until a University of Melbourne article in 1944.

Cranbourne occasionally popped up in the popular press. *The Strand Magazine* was an illustrated monthly first published in London in 1891, and which ran until 1950. It targeted the mass market, and for sixpence readers enjoyed short-form fiction and articles of general interest. It also serialised popular works from Arthur Conan Doyle and, later, Agatha Christie and P. G. Wodehouse. In 1896 one of its contributors, William G. FitzGerald, wrote a piece entitled 'The Romance of the Museums,' in which he described articles on display in the British Museum and the Natural History Museum, and gave interesting back-stories to these pieces. Along with tales of Marie

Antoinette's toilet table, an Egyptian mummy, and a broken piece of jasper from Cairo which in cross-section displayed a portrait of Chaucer, FitzGerald discussed Cranbourne No. 1 – 'an immense meteorite.' It is one of the few accurate statements in this short section, to wit: James Bruce found the meteorite, it was shown in the exhibition of 1861, Bruce arranged for its transport to England, and the Victorians discussed sending a ship in pursuit to retrieve the specimen. In addition to these falsehoods FitzGerald stated, rather unkindly, 'The Melbourne Museum, however, continued to clamour childishly for its meteorite.'[15]

* * *

The Victorian gold rushes brought a population influx on a scale not repeated for over one hundred years, until the immigration waves after World War Two. They also seeded the civic, cultural, and scientific institutions in the colony with their future leaders. Many of those involved in the Cranbourne affair had journeyed south to seek their fortune. They came for the gold but then stayed, mostly, and helped to build a colony.

Hugh McKay farmed his Sherwood parish property up to his death in December 1875. The seven-room house in which he and Jennet lived was also home for a niece and nephew, Elizabeth and James, children of Jennet's younger sister Robina. Their mother had married the year after the doleful arrival of the *Glen Huntly*, and gave birth to four children in six years. She died while in confinement with her fifth, and the apparently childless Hugh and Jennet cared for the two siblings thereafter. Hugh's will passed his 650 acres to Jennet; upon her death it would be apportioned between Elizabeth and James.

James Lineham died in April 1901, and lies in the Cranbourne cemetery. In 1878 he bought another property, at Cardinia, living there for 10 or 12 years before he and Charlotte moved back to their Clyde holding and built a wooden house. In 1971 descendant Jim Lineham was photographed standing in the corrugated rows of the field where No. 2 was unearthed, by a sapling planted to mark the spot. Suburbia has now encroached; a sporting field and recreation reserve, fittingly named Lineham Oval, occupy the land where the homestead once stood and the market garden furrows are overlaid with brick and bitumen.

James Bruce did not return to Victoria. He had first advertised his Cranbourne properties for let in mid-1862, prior to his departure, and again in 1865. Eleven years later 'Sherwood Park' was put up for auction as a 'first-class grazing estate.' Bruce lived in Chislehurst, south-east of London, for several years, building a mansion named Black Mount there in 1872. He died in Inverquhomery, Scotland, in February 1900.

Augustus Abel remained in Ballarat and continued as a mineral dealer. He was active in the Mechanics Institute, to which he donated part of his collection – later transferred to the School of Mines. He lived to 80, pre-deceased by his wife and only son, feeble and eccentric. Afflicted with neuralgia and subject to 'delusion' in his final years, he required the care of neighbours for his daily wants. A dispute simmered over his estate, with his will in favour of his surviving sisters in Germany contested by a Ballarat woman who claimed a bequest in her favour via a different document. The judge in the Equity Court found the deceased 'in some things to have been much subject to her dictation.'

Edmund FitzGibbon retained his Town Clerk position until 1891, a role in which he resolutely championed the public ownership of Melbourne's civic infrastructure and preservation of its parklands. Perhaps his compunction over the fate of No. 1 drew him to view its display on visits to the British Museum in 1876 and, in 1892, the Natural History Museum. He was appointed first chairman of the Melbourne and Metropolitan Board of Works, the body assembled to oversee sewage and water supply to the city. He remained in the position until his death in 1905. His statute stands in St Kilda Road; erected in 1908 by a grateful council, it presents him in bronze on a granite pedestal, gazing across the Yarra in appraisal of the city he served for most of his adult life.

After Sir Henry Barkly's tenure in Victoria, in 1863 he transferred to Mauritius and served as governor until 1870. He was then posted to the Cape of Good Hope as Governor of Cape Colony. However, the conduct of his co-duties as British High Commissioner in South Africa led to friction over policy with the Secretary of State for the Colonies, the Earl of Carnarvon. It was a regrettable finale to an otherwise commendable colonial career, but even so, after his recall he was appointed to the royal commission on colonial defence, in 1879, under Carnarvon's chairmanship. Barkly continued his intellectual and scientific recreations in retirement.

Georg Neumayer returned to Germany in 1864 and sustained his numerous scientific pursuits, oceanography and meteorology among them, over a long and decorated career. In 1869 the report of his extensive Victorian survey was published. Appropriately for a magnetician, he maintained a keen interest in the Earth's polar regions, fostering several expeditions in this field and supplying a vigorous impetus to the heroic age of Antarctic exploration. Roald Amundsen studied under him. Ennobled in 1900, the dashing *enfant terrible* matured into the grand old man of German science, in the tradition of Humboldt and Haidinger.

Alfred Selwyn added to his impressive body of work with the Geological Survey of Victoria, conducting field trips and generating maps of the highest standard, until the government abruptly disbanded his department in 1869. Disillusioned, he left Australia that year and took the post of director with the Geological Survey of Canada. He spent the next 25 years investigating and mapping the country's geological landscape, through often arduous field work. In 1884 he sampled ore-bearing rocks from cuttings made for the Canadian Pacific Railway near Sudbury, Ontario. The surrounding basin was later identified as one of the largest impact craters in the world. Formed by a meteoric, or possibly cometary, collision so powerful it pierced the Earth's mantle 1.8 billion years ago, it has rich deposits of nickel and copper – ore derived from mantle magma that filled the massive crater. Selwyn finished his days in British Columbia, seeing in the new century retired and content, his CMG assured.

Robert Brough Smyth led the Department of Mines until 1876, and maintained a steady feud with Selwyn for most of the 1860s. In contention was the level of assistance the Survey provided to the mining fraternity, the practicality of Selwyn's maps and the speed of their delivery. Smyth outmanoeuvred Selwyn by agitating successfully for the Survey's disbandment in

1869 and then leading the reinstated organisation from 1870. But he eventually fell foul of his own intemperate management style. Officers of his department levelled charges of 'tyrannical and overbearing conduct' against him in 1876. Despite Frederick McCoy's support, and recognition of his 'unremitting energy and zealous labours in the public service,'[16] the government board that examined the issue substantiated the allegations. Smyth resigned his public offices but remained on the Aborigines Protection Board, in 1878 publishing a valuable two-volume work, *The Aborigines of Victoria*.

Richard Daintree had the fortunate ability to form friendships wherever he went. From 1864 he established himself in north Queensland and prospected gold mines, acted as government geologist, and was later appointed the colony's agent-general in London. His photographic skills served to promote his adopted home in the many exhibitions he attended. He died of tuberculosis in Kent in 1878. His friend, explorer George Dalrymple, named the Daintree River of north Queensland after him, the title also given to the spectacular adjacent rainforest.

John Macadam did not survive the year of No. 1's departure for Britain. He died in September of pleurisy complications while on the way to New Zealand to give evidence in a murder trial. His was a hectic and eventful ten years in the colony. George Foord enjoyed greater longevity, and published widely on chemistry and its application in scientific journals and the Royal Society's transactions.

* * *

What did the removal of Cranbourne No. 1 mean to Victorian colonists? It was a lost opportunity to identify with a notable and intriguing local artefact. One which communicated the exceptionalism of its location in the manner of Victoria's stupendous gold yield. The yellow metal had appeared as nuggets and alluvial traces for years and was now being loosened from quartz veins deep beneath the surface by powerful crushing machines. Familiarity, while not quite breeding contempt, had enabled a level of 'gold fatigue' by the mid-1860s, in contrast to the excitement of the early 1850s. But as lustrous and valuable as gold was, it did not convey the otherworldliness, nor the downright peculiarity, of a space-borne visitant.

The loss was keenly felt in some quarters. In 1869 Robert Brough Smyth authored *The Gold Fields and Mineral Districts of Victoria*, in which he documented the geological landscape and mining history of the colony. Among its pages he grieved the forfeiture of Bruce's iron; 'the larger mass is now in the British Museum – an object of vulgar curiosity,' and reflected on the lessons taken from the struggle:

> *That a meteorite of great size, of surpassing interest – and evidently exhibiting peculiarities of composition, if not of structure – should have been allowed to pass away from us as a thing of no value, is, I am bound to say, not creditable to us as colonists. Such an event we may rest assured can never happen again; and though it might be scarcely worthwhile to reclaim it, it is not beyond possibility that our great meteorite may yet again be re-purchased and added to the National Collection.*[17]

Of known space rocks, the trans-Murray Baratta specimen, initially identified in 1845, was not investigated in detail until decades later. During the interregnum Cranbourne was the only local meteorite acknowledged as such. Since it was also the largest iron specimen in the world, it served to further enhance the colony's premier mineralogical status. Victorians could rightfully take pride in the results of their industry in capturing the wealth of their natural resources and turning this to their advantage. Cranbourne No. 1 crowned the terrestrial blessings of the colony with something truly awe-inspiring. It would be all the poorer for its loss. Smyth would hope in vain.

Should it be returned, and under what arrangements might that even happen? The Melbourne Museum is the obvious destination, where it would be united with No. 2 for the first time since their fiery separation, high in Earth's atmosphere. But it seems Victorians will continue to rue, for some time yet, the outcome of the 19th century turf war conducted over their meteorite, and the all-important relinquishing of home ground advantage: possession.

12
Epilogue

Even less do they see the meteor that cuts open the night sky
and instantly sends out a brilliance more beautiful than the day.
Yu Ouyang Untitled

Meteorites like Murchison, with a range of organic compounds in their makeup, contribute to conjecture about life on Earth being seeded from space. But in rare cases these space travellers have interacted with the petrification of long-dead marine organisms. In the same way as relics of ancient life forms are encased and preserved in sedimentary rock, some meteorites fall in locations conducive to fossilisation. The principal cluster of such specimens is in southern Sweden, in a working limestone quarry. Here the stone is a compressed calcareous residue of Ordovician-era marine life; cephalopods and trilobites.

The Ordovician was a 42-million-year transition classification that bridged the contentious Cambrian and Silurian periods of Adam Sedgwick and Roderick Murchison, and was book-ended by conspicuous extinction events. During this time, starting 485 million years ago, the Gondwana supercontinent dominated the antipodes. North of the equator was virtually all ocean, and a steady deposition of arthropod exoskeletons on the ocean floor laid down, very slowly, what would become the limestone beds of the Thorsberg quarry.

The Ordovician also saw Earth bombarded by chondritic meteoroids. This heightened flux is thought to originate in a collision between two bodies of the Asteroid Belt 470 million years ago. One of these is the source of the Thorsberg meteorite fossils: astronomical objects that journeyed to Earth, fell into the northern ocean, sunk to the sea floor, and were subsumed by sedimentary rock-making action. They form a fascinating palimpsest; antique vestiges of the primitive Solar System, fossilised twice over.

To add to the peculiarity of these interpolated Thorsberg samples, they have meteoric siblings elsewhere in the world. In March 2003 a fiery meteor shower lit up the Chicago suburb of Park Forest. The collected meteorites were identified as coming from the same asteroid collision as those found at Thorsberg, based on their similar signatures of radiogenic gases – minute traces of cosmogenic isotopes of helium and argon.

There is one more twist. Cosmogenic isotopes are created by the effect of cosmic rays when an asteroid is broken up in collision and its interior is exposed to the radiation. Once a meteor enters Earth's protective atmosphere the process ceases. This means the amounts of these isotopes measured in a meteorite can help discern the period it spent travelling in space. Intriguingly, the oldest of Thorsberg's fossil meteorites — those found in the deepest layers of limestone — reflect a shorter space-travel time than that of related samples with a younger terrestrial age. Jupiter's gravitational puppeteering may well be the cause. An enormous impact, deep in space, pelted rocks at Earth via a cosmic conveyor belt for two million years, and continues to do so today, albeit at a reduced rate. Its first arrivals (Thorsberg), took about 50,000 years to make the journey, possibly hastened by a timely Jovian resonance close to the massive collision; the later visitors (Park Forest), approximately 470 million years.

Maurie Richardson is an affable retiree whose home is in Langwarrin, a south-eastern suburb of Melbourne near the south-western conclusion of the Cranbourne shower's strewn field. His early working career was spent in engineering, on major projects. After re-training he became a teacher, then administrator, with the Victorian Education Department. Music has been a long-held passion. Life is now more leisurely, but he remains an active chorister and performer.

In 2008 Richardson was acting principal of the small primary school at Clyde, a rural hamlet six kilometres east of Cranbourne. Alexander Patterson's St Germains run, on the fringe of the Koo Wee Rup Swamp, once covered this land. It is a market garden area where local farmers till sandy loam soils to produce vegetables for the Melbourne market. A poultry feed mill was set up in the 1970s, and its high structure towers over the surrounding patchwork of vegetable rows.

By May of 2008 a NASA-led consortium of international universities had launched the Phoenix robotic spacecraft and placed its lander on the surface of Mars. The International Space Station received multiple space shuttle visits as its construction neared completion. Space was in the news. To leverage student interest in these events, in the second term of the year Clyde Primary promoted a whole-of-school theme on space, conducting several space-related activities and assigning course work to students. Then a fortuitous event took the project in a new and exciting direction.

In April, Clyde farmer Glenn Blundy decided to remove a troublesome rock from the property on which he worked. After years of ploughing and crop rotation in this section of the holding – 160 acres on which cabbages and potatoes were grown and store cattle grazed – the small piece lay near a fence line on the eastern boundary of its paddock. But a passing machine had recently clipped it and it was now sitting in an inconvenient spot; Blundy's tractor often collided with the protuberance. So he manhandled it into his utility vehicle, then brought it over to one of the farm sheds. Working at the same farm was Gerard Sadler, son-in-law of the property owners and president of the Clyde Primary school council. He helped Blundy move the rock from the ute and into a bin in which various wood scraps and other refuse was held. Sadler noticed the mass had a russet colouring and approximate dimensions of a shoebox, but when placed on scales – normally used to weigh pumpkins – it registered an extraordinary 85 kilograms. It was different in appearance and heft to the coffee rock found in the area, and which often caught on ploughshares during tilling.

These characteristics puzzled Sadler. The sample had rust-like patches. The field from which it was retrieved held a catchment dam, and was often flooded. The rock would have been submerged for prolonged periods. Sadler thought it might be ferrous slag of some sort. He decided to sample a small piece but, busy with farm duties, several weeks passed before he returned to the task. He first used a pressure washer to clean the rock's exterior. Then he applied an electric angle grinder to shear off a section, taking a surprising twenty minutes to do so and destroying a ten-inch blade in the process. On the clean face then exposed, smooth and metallic, he discerned the faintest hint of striation, something he compared to the large-scale basalt columns seen at Keilor's Organ Pipes national park, above Jacksons Creek on the other side of Melbourne.

At a subsequent school council meeting, while in conversation with Richardson about the space project, Sadler mentioned this unusually heavy rock. The assistant principal was immediately interested. His engineering studies had included units in geology and material sciences, and he knew of the earlier meteoritic finds in the district – Clyde Primary is scant kilometres from Cranbourne No. 1's landfall and even closer to where No. 2 was unearthed. He had an inkling that the rock could be a related find. Richardson asked if it might be brought to school for the students to view. Sadler agreed but requested the find's location be kept secret, not wishing potential disruption to farming activity nor damage to property.

When the rock arrived at the school Richardson noted the newly cleaned surface was already showing fresh signs of rust. It was placed in the school library, on a low table where small flakes of its rufous outer crust soon sprinkled the cloth doily placed underneath it. Richardson sent Sadler's sample piece and some photographs to the Melbourne Museum Discovery Centre with a request for testing and evaluation. In the meantime, the children had an interesting new subject to examine. The question was posed: is this rock possibly a meteorite? Students tested the sample with magnets, conducted activities involving measuring the specimen against their own body weight, and drew pictures of meteors falling to earth. Their laminated drawings formed a display around the sample, and included innocent remarks such as 'This is a heavy rock. I think it has fallen from out of space' and 'The rock has got black marks and it has red too.'

After a few months Sadler contacted Richardson seeking an update. With no news to report, the assistant principal inquired at the museum, where he learned that numerous exhibitions had held up the examination of the specimen. Many such requests are made of museum staff, and in almost all cases a non-cosmic source is determined for the supplied samples. But Richardson then received an excited call-back from a Melbourne Museum representative in August. He reported that an electron microscope was used to examine the sample, and substantial iron and nickel were present in matrix, along with troilite, an iron sulphide. It was good news. Their rusty rock was indeed a meteorite, and it very likely belonged to the Cranbourne shower. Richardson made the announcement over the school's public address system that day, to cheers of jubilation from the students. By now they had dubbed their pet project 'Clyde.'

The *Cranbourne Star* newspaper ran a story on the find, and included a photograph of two curious preparatory grade students posing with the meteorite and holding magnet and tape measure, making their observations. The article fostered talk of the meteorite's value, and with some large sums being bandied about the school council decided that it should be returned to the farm for safekeeping. Sadler ferried it back to the shed and it was hidden in plain sight among the farm equipment and various odds and ends. The property owners Bill and Merle Blundy wished to donate their find to Museum Victoria, an offer gratefully accepted by the institution. When time came for the transfer, Sadler again brought it to the school where museum geologist and manager of natural science collections Dermot Henry took possession after a presentation to the students. Merle and two of her granddaughters, pupils at the school, posed for photographs with Henry and the fragment before it was taken away to its new home.

In early December Cranbourne No. 13 was the centrepiece of Museum Victoria's exhibition 'Earth Quest – Outer Space to Inner Earth' at the Scienceworks site in Yarraville. Clyde Primary students were in attendance at the unveiling of their 'Clyde.' Another participant was Merle Blundy, accompanied by her daughter-in-law. A museum driver especially chauffeured them to the event. Work duties kept Sadler and Glenn Blundy from attending.

Richardson is an engaging individual with an easy rapport, and was popular with his students. He draws upon a convict ancestry, both paternal and maternal, and has a charmingly roguish manner one can readily attribute to his Tasmanian prisoner forebears. He tells a yarn about the famous meteorite fragment with a glint in his eye. When placing the specimen on that sturdy coffee table in the school library, he mishandled it and it fell, striking his foot a glancing blow. It was not a serious injury, but Education Department rules required that an incident report be filed. Its description field noted the following: 'Hit on foot by falling meteorite.'

* * *

Early in the Solar System's development, a rampant planetesimal shattered upon colliding with another cosmic body. Its debris then rode the Solar System's gravitational ebbs and flows, possibly for billions of years, before one piece – a fragment of the differentiated planetesimal's core – underwent a final perturbation and was slung at Earth, to breach the atmosphere somewhere over New South Wales or the Pacific Ocean. This passage violently reduced the meteoroid's cosmic speed and ablated its surface with temperatures near 2,000 degrees Celsius. Unable to withstand the intensity of the incursion, it broke apart high above the Australian land mass and finally fell to earth, riveting a long chain of iron fragments into the damp plains of Mar-ne-bek.

Thirteen meteorites have been uncovered from this stippled corridor since 1854. Are there more out there, waiting to be found? Quite possibly. Strewn fields usually fashion an elliptical shape for falls not directly vertical, and the smaller members of a meteorite shower generally lag behind the larger.[1] In Cranbourne's case this implies a north-east to south-west fall direction for the narrow field. So the wooded gullies north-east of Beaconsfield could harbour a number of specimens, but the country south-west of Langwarrin and Pearcedale, with its well-developed housing areas, may be played out.

With three exceptions, including that of the premier fragment, the main Cranbourne meteorites are housed in Australian collections. Many smaller samples have been variously sawn, hacked, or otherwise excised from among these and are distributed worldwide. Of the specimens that came to light in the 1850s, Cranbourne No. 1 remains in London and No. 2 is on display at the Melbourne Museum. Hugh McKay gave the No. 3 fragment, found close to No. 1 and probably thrown off during the larger mass's descent, to Edmund FitzGibbon, but it was eventually lost.

Twenty years elapsed before the next Cranbourne find. In 1876 the specimen to be allocated No. 9 turned up in a cutting east of the Beaconsfield train station, during construction of the Gippsland railway. A blacksmith made the discovery and, thinking it part of a mineral vein, sent a piece to the Victoria government geologist, Reginald Murray, who surmised its origin. A European buyer purchased the 75-kilogram meteorite and in time a Bonn mineral dealer acquired it, after which it was cut up and widely dispersed. The German mineralogist Emil Cohen analysed a sample, and among its minerals he noted an iron carbide compound, presenting as fine crystal rods and already named after him – cohenite.

In 1886 the No. 10 fragment was revealed on land to the south-east of Langwarrin. An Alfred H. Padley was reported as the owner of the property but in fact it was held by the Cosmopolitan Land and Banking Company, one of many such organisations that sprang up in Melbourne during the land boom of the 1880s. The company had a considerable holding in Langwarrin which was subdivided and sold on to owner/occupiers, investors, and speculators – a practice that helped fuel the rampant land price increases of this period. Padley was Cosmopolitan's managing director. In a letter to *The Age* more than 40 years after the event he described No. 10's discovery, made by the Langwarrin Estate's overseer when ploughing in an orchard. Upon receiving news of the find Padley gave instructions for the mass to be dug up, a task which required a team of sixteen bullocks – the meteorite weighed over 900 kilograms. Reginald Murray was invited to inspect the discovery on site, and confirmed it 'a meteorite of the greatest scientific value.' One can assume Murray suggested Padley's next course of action, for the find was offered whole to the Industrial and Technological Museum, later part of Frederick McCoy's National Museum of Victoria. The bullocks were again deployed, this time to cart the prize to Frankston, from where a train carried it to Melbourne.

In his 1927 letter Padley referred to Walcott's 1915 study of the Cranbourne samples, and how he had expected more 'scientific questions' and 'further information' on these meteorites to be circulated as a result. Furthermore, no concerted attempts had been made to uncover other meteorites from the area and, had he offered his specimen to Germany, there would have been 'searching scientific work' as an outcome. Also, he would have obtained 'a large sum of money for it.' He seemed a little piqued, even after all that time. But he had a point. With the identification of No. 10, and the known placement of the No. 9 'Beaconsfield' sample identified 13 miles to the north-east, the lengthy and linear nature of the Cranbourne strewn field became apparent. So those curious enough to look for more meteorites now had a reasonable frame of reference. By coincidence, as a boy Padley had witnessed a meteorite fall to the north-west of Melbourne in 1857/1858, when a 'bolt from the blue' resulted in a meteorite 'which somewhat resembled a cannonball' being un-earthed.[2]

Preceding the turn of the twentieth century, a severe bushfire scorched through South Gippsland and Western Port. The 'Red Tuesday' fire and associated blazes of early February 1898 claimed twelve lives and destroyed much grass, gardens, and fencing between Tooradin and Cranbourne. The pine plantations of the Sherwood area were eliminated, and although most homesteads were saved newspapers reported properties in the area being 'swept clean' of outhouses, crops, and orchids.[3] In 1903 this fire-hardened country surrendered another meteorite fragment. The No. 11 iron was found north-west of Pearcedale, but not revealed publicly until much later, in 1938, when it was sold to the Smithsonian Institute. Weighing an impressive 760 kg, it was the fourth heaviest Cranbourne specimen to date and like many of the others it lay just below ground level. Confusingly, No. 10 (long called the Langwarrin meteorite) is located closer to the modern Pearcedale than No. 11, which takes the name Pearcedale but whose find site is nearer to today's Langwarrin. The village of Pearcedale is named after the family that bought up much of the allotments of Cosmopolitan's Langwarrin Estate, including its small commercial area, in the 1890s after the crash in property prices that followed the land boom.

The year 1923 was a highlight for Cranbourne finds. On a plot just north of Hugh McKay's old holding, farm manager George Bacon unearthed a 1,270 kg fragment that would become No. 4. Bacon oversaw the now-reduced Sherwood Park, earlier subdivided by its entrepreneur owner Hans Irvine, the recently deceased and hitherto 'wine king of Australia,' famous for his Great Western vineyards.

Later in the year a farmer's plough again proved an efficient meteorite detection device. A. R. Croker worked allotment 33 in Sherwood, another section of James Bruce's former estate and which shared its south-western corner border with the old McKay property. When furrowing a paddock he turned up three specimens in close proximity. These were allocated numbers 5, 7, and 8. Stanley Hunter of the Mines Department arranged for their purchase, and posed for a newspaper photograph in early 1924 with three hefty-looking specimens at the state store in South Melbourne.

In a similar manner to No. 11, what became No. 12, a 23 kg fragment, was first found in 1927 north of Pearcedale but not identified as meteoritic until 1982. Then in 1928 the 40.5 kg No. 6 fragment was brought to light during earthmoving works on the Princes Highway west of Pakenham. It would be another 80 years before the students at Clyde Primary learned of the No. 13 fragment's inclusion in the Cranbourne roll call.

The length of the Cranbourne strewn field suggests a shallow-angle atmospheric entry by the parent meteor. As well as its near arrow-straight array, a clustering effect is noticeable within the field. Four such groupings occur; at Pakenham/Officer (fragments 6 and 9), Clyde (2, 13), Cranbourne South (1, 3, 4), and Pearcedale/Langwarrin (10, 11), and these indicate a series of aerodynamic separations of the meteor at altitude. It would have been quite a sound-and-light show. Each cluster, being a scatter ellipse or mini strewn field, has at least one lighter specimen and a heavier down-range fragment, supporting the proposal that weightier bodies' greater momentum carries them further than their smaller relatives; the lighter fragments have a greater surface-area-to-mass ratio, and so decelerate first and fall to the ground at the leading edge of the ellipse.[4]

But Cranbourne No. 1 was the prize fragment, and reigned as Australia's foremost meteorite for 100 years. Then, in 1966 on the Roe Plains – a strip of marine dune country between the Nullarbor Plain and the Great Australian Bight in the far south-east of Western Australia – a 12-tonne monster was identified near Mundrabilla. That mass, another iron octahedrite, holds pride of place in the Western Australia Museum's collection in Perth.

So the 1960s saw Cranbourne No. 1 knocked from its pedestal, first by the massive Mundrabilla and then by the noisy interjection of Murchison and its intriguing chemistry. Not as big as the former, nor as interesting as the latter, Cranbourne still holds celebrity status in the Australian meteoritic pantheon – albeit with a whimsical appellation. It will always be the one that got away.

* * *

On the Berwick-Cranbourne road approaching Cranbourne township is a large aquatic centre, with nearby library, basketball courts and indoor skate park. It's the sort of community complex prized by city councils across Melbourne, and the swimming pool structure is large and modern. Out front is a sizeable artistic installation of unusual subject matter. Comprising three curved trees rendered in steel, representing eucalypts, on a raised platform above a field of sand, it's a stylised wetland with Zen garden overtones, in which eight red-brown rocks are embedded. These represent the main Cranbourne meteorite pieces uncovered since 1854, a strewn field in cast-iron abstract. There's No. 1 as a bulky ochre menhir, with its various satellite stones in approximation of their found locations. They rise out of the ground like a field of icebergs from water, suggesting greater mass beneath.

They are perhaps given a passing glance by families and clamouring children as they to-and-fro by the inside pool and gymnasium, and some may even stop to read the small plaque on its rusty plinth, but most pass it by. Although a sleek rendition of a worthy local celebrity, perhaps visitors consider it a little trite. But it's an improvement upon earlier council efforts to symbolise the town's celebrated meteorite; a rock ensemble hung in a scaffold arrangement, an outsized nursery mobile, which stood in view of passing traffic for many years in a park by the South Gippsland Highway.

***

The Natural History Museum in London comprises an imposing Romanesque edifice that fronts a lengthy sweep of Cromwell Road south of Royal Albert Hall. I visit on a cooling April afternoon, and admire the building's exterior façade and internal galleries' use of terracotta tiles, in pale blue and beige, made to marked effect. Inside, Cranbourne No. 1's final resting place is a nitrogen-filled glass display case, in its eponymous boutique, on the ground level. The setting is stark; a pale wooden floor and high, dark-painted ceiling with exposed pipes and lighting rails; all thin metallic shelving and rigid right angles. This is a minimalist geometric warehouse; part curio display, part salon, part gift shop. Patrons bend over encased crystals and other curiosities, or rifle lethargically through racks of merchandise. School children chatter.

The meteorite is majestic and forlorn in equal measure. Few visitors pause to examine its lumpen metallic-brown mass or read its display case's text. Not that there is much detail imparted thereon; weight, date of discovery, and its premier position among the museum's 5,000+ meteoric samples. Some generic meteorite information. No mention of the protracted tug-of-war that led to the specimen's placement in this Kensington souvenir stand. A catalogue number, 55532, is neatly stencilled in white numerals on the rusty surface near two rough-hewn edges, probably the remains of protuberances excised those many years ago. A more precise, and much larger, excision is visible on the shoulder of the piece. This non-weathered shearing shows an etched grey face, with the octahedrite's Widmanstatten patterns vaguely apparent.

It's all rather odd, this juxtaposition of interplanetary object and retail insouciance. Our meteorite is not paid its due respect by such placement. Above all it is devoid of backdrop; without reference to Western Port's ruddy complexion, or its windy panorama, or the grey-green patina of the bush. It's absent any indigenous association or placement within country. Sadly, it fulfils the worst fears of McCoy and others whose pessimistic predictions ring righteously across a century-and-a-half. It should not be a trophy, reinforcing outdated notions of empire and annexation, but a celebrated motif of place, people, culture, and connection.

The boutique staff are pleasant and eager to assist but have little detail to offer on their once-famous specimen. The books in their collection do not even reference Cranbourne. I take my notes and pictures, then take my leave, and step out into the chilly London gloaming. A wistful thought plays across my mind; how fitting it would be to exit my viewing into a Victorian setting, to hear the buzzing song of cicadas and chime of bellbirds, to sense the wafting aroma of eucalyptus, and know that Cranbourne was once again united with the place, the earth, the *country*, in which it was cradled for so long.

Notes

Chapter 1
1 Cassell's Illustrated Family Paper, vol. 1, no.6., London, Saturday, February 4, 1854.
2 1851 'MEETING TO OFFER A REWARD FOR FINDING A GOLD MINE', *The Argus* (Melbourne, Vic.: 1848–1957), 11 June, p. 4, viewed 7 May 2020, http://nla.gov.au/nla.news-article4778440.
3 1953 'BLACK' THURSDAY, 1851', *Portland Guardian* (Vic.: 1876–1953), 12 January, p. 2. (MIDDAY), viewed 05 May 2020, http://nla.gov.au/nla.news-article64433695.
4 *The Making of a Governor,* Dianne Reilly Drury.
5 *Official Catalogue of the Melbourne Exhibition 184.*
6 'Gipps Land Guardian', *Gippsland Guardian* (Vic. : 1855–1868), 9 September 1859: 2. Web. 19 April 2019 <http://nla.gov.au/nla.news-article112489234>.
7 *The Good Country. Cranbourne Shire,* Niel Gunson 1968.
8 Ibid.
9 *Results of the Magnetic Survey of the Colony of Victoria executed during the years 1858–1864,* Neumayer, G., 1869.
10 *Descriptions of the Victorian Meteorites, with notes on obsidianites. By R. Henry Walcott, F.G.S., Curator of the Geological and Ethnological Collections.*
11 *Natural History Museum, Meteorite letters A-G (Cranbourne)Sir Henry Barkly.*
12 *Ferdinand Mueller and the Royal Society of Victoria,* R.W. Home, CSIRO Publishing.

Chapter 2
1 Transactions of the Royal Society of Victoria 1860, vol. 5.
2 Ibid.
3 *Experiments on the Stability of FeOOH on the Surface of the Moon,* Lawrence A. Taylor and Jacqueline C. Burton.
4 1896 'Concerning Meteors', *The Herald* (Melbourne, Vic.: 1861–1954), 13 February, p. 3, viewed 22 April 2019, http://nla.gov.au/nla.news-article241278373.
5 *Influences of German Science and Scientists on Melbourne Observatory,* Barry A.J. Clark, CSIRO Publishing.

6   *Three Expeditions in the Interior of Eastern Australia* Mitchell, Thomas Livingstone, 1839.

7   *Results of the Magnetic Survey of the Colony of Victoria executed during the years 1858–1864,* Neumayer, G., 1869.

8   Ibid.

9   *Pakenham Gazette,* 3 October 2001.

10 *Results of the Magnetic Survey of the Colony of Victoria executed during the years 1858–1864,* Neumayer, G., 1869.

Chapter 3

1   1876 'Advertising', *The Argus* (Melbourne, Vic.: 1848–1957), 27 May, p. 3, viewed 22 April 2019, http://nla.gov.au/nla.news-article7440171.

2   *Catalogue of the Victorian Exhibition 1861.*

3   Ibid.

4   Ibid.

5   *Bearbrass: Imagining Early Melbourne,* Robyn Annear, 2015.

6   'Reminiscences of John Pascoe Fawkner', *La Trobe Library Journal,* no. 3, April 1969.

7   1861 'THE GREAT METEOR AT VICTORIAN EXHIBITION 1861', *The Herald* (Melbourne, Vic.: 1861–1954), 12 October, p. 5, viewed 8 August 2021, http://nla.gov.au/nla.news-article244247731.

8   Annotated note 'Mr Abel's note to Prof Abel of Woolwich'. Meteorite letters. Natural History Museum.

9   *Ferdinand Mueller and Charles L Trobe,* Tom Darragh. La Trobeana – vol. 2, no. 2, June 2004.

Chapter 4

1   *Australia's Great Comet Hunter,* Dr Ragbir Bhathal (https://academic.oup.com/astrogeo/article-abstract/51/1/1.23/199612).

2   *Letter, Murchison to Maskelyne November 20th, 1861,* Natural History Museum (NHM).

3   *Letter, F. Abel to Maskelyne,* 9 December 1861, NHM.

4   *Letter, McCoy to A. Abel,* 16 December 1861, Outward letter book, Melbourne Museum.

5   *The Victorian Naturalist,* vol. 118(5), 2001.

6   Ibid.

7   Ibid.

8   *Catalogue of the Victorian Exhibition 1861.*

9   *The Victorian Naturalist,* vol. 118(5), 2001.

10 *Formation of Museums in Victoria,* Frederick McCoy, 1856.

11 *Colonial pride and metropolitan expectations: The British Museum and Melbourne's meteorites.* Lucas, A.M.; Lucas, Paula, Darragh, T. A.; and Maroske, S. B*ritish Journal for the History of Science,* vol. 27, 1994.

Chapter 5

1 *Letter, McCoy to J. Bruce*, 3 January 1862, Outward letter book, Melbourne Museum.

2 Victoria. Governor (1856–1863: Barkly), Roderick Murchison, Alfred Richard Cecil Selwyn, and Frederick Arthur Stanley. *Geological Survey: Return to an Address of the Legislative Assembly Dated the 9th February 1860 for – A Copy of His Excellency's Despatch to Lord Stanley of the 12th July 1858, Together With the Report of the Geological Surveyor Mr Selwyn of the 13th July 1858.* Melbourne: John Ferres, Government Printer, 1860.

3 Ibid.

4 *Letter, McCoy to J. Bruce*, 3 January 1862, Outward letter book, Melbourne Museum.

5 Ibid.

6 Letter, J. Bruce to *The Argus*, 3 December 1862.

7 Ibid.

8 *Impressions of Australia Felix During Four Years Residence in that Colony,* Richard Howitt.

9 *A Summer at Port Phillip,* Robert Dundas Murray.

10 Ibid.

11 *Letter, F. Mueller to Maskelyne*, 25 January 1862, NHM.

12 *Letter, McCoy to J. Bruce*, 4 February 1862, Outward letter book, Melbourne Museum.

13 *Letter, Bruce to McCoy,* 9 February 1862, NHM.

14 Ibid.

15 *Letter, McCoy to J. Bruce*, 13 February 1862, Outward letter book, Melbourne Museum.

16 Ibid.

17 Ibid.

18 *Letter, Bruce to McCoy, 1*3 February 1862, NHM.

19 *Letter, Mueller to McCoy,* 14 February 1862, NHM.

20 *Kew: The History of the Royal Botanic Gardens,* Ray Desmond.

21 *Ferdinand Mueller and the Royal Society of Victoria,* R.W. Home, CSIRO Publishing.

22 Letter, Mueller to Joseph Hooker, 24 October 1865 – Mueller, Ferdinand

von, 1825–1896 & Home, Roderick Weir. 1998, *Regardfully yours*: selected correspondence of Ferdinand von Mueller / edited by R.W. Home ... [et al.]. Peter Lang Bern; New York.

Chapter 6

1. Bevan, A. & Bindon, Peter. (1996). Australian Aborigines and Meteorites. Records of the Western Australian Museum, vol. 18, pp. 93–101.
2. Letter, McCoy to Mueller, 14 February 1862, NHM.
3. Letter, F. Mueller to Maskelyne, 20 February 1862, NHM.
4. *Letter, McCoy to Bruce,* 21 February 1862, Outward letter book, Melbourne Museum.
5. *Formation of Museums in Victoria,* Frederick McCoy, 1856.
6. *The Victorian Naturalist,* vol. 118(5), 2001, Sir Frederick McCoy FRS – an Overview Malcom Carkeek.
7. *The Victorian Naturalist,* vol. 118(5), 2001, Professor Frederick McCoy and the National Museum of Victoria, 1856–1899 Carolyn Rasmussen.
8. Descriptions of the Victorian Meteorites, with Notes on Obsidianites Henry R. Walcott.
9. *Results of the Magnetic Survey of the Colony of Victoria executed during the years 1858–1864,* Neumayer, G., 1869.
10. http://nla.gov.au/nla.news-title1190 (Trove).
11. *Letter, Bruce to brother,* 21 March 1862.
12. *Letter, Foord to Selwyn,* 27 February 1862, NHM.
13. *Letter, Selwyn to Mueller,* 28 February 1862, NHM.
14. *Ferdinand Mueller and the Royal Society of Victoria,* R.W. Home, CSIRO Publishing.
15. *Letter, Mueller to Maskelyne,* 25 March 1862, NHM.
16. *Anniversary address, 1862* Sir Henry Barkly Transactions of the Royal Society of Victoria, 1861 to 1864.

Chapter 7

1. Bevan, A. & Bindon, Peter. (1996). Australian Aborigines and Meteorites. Records of the Western Australian Museum, vol. 18, pp. 93–101.
2. The International Exhibition of 1862: the illustrated catalogue of the Industrial Department. https://archive.org/details/internationalexh01lond/page/n121.
3. Trove – Melbourne Punch Thu 11 September 1862. https://trove.nla.gov.au/newspaper/article/174527876?searchTerm=Punch%20exhibition%201862&searchLimits=l-state=Victoria||||l-decade=186.
4. *Letter, R Brough Smyth to McCoy,* 15 April 1862, Melbourne Museum.
5. *Letter, R Brough Smyth to McCoy,* 28 July 1862, Melbourne Museum.

6   *Letter, N. Story-Maskelyne to Owen*, 21 May 1862, Transactions of the Royal Society of Victoria, 1861 to 1864.
7   *Letter R. Murchison to Sir Henry Barkly*, 22 May 1862, Transactions of the Royal Society of Victoria, 1861 to 1864.
8   1862 'VICTORIA', *Empire* (Sydney, NSW: 1850–1875), 12 September, p. 2, viewed 4 November 2019, http://nla.gov.au/nla.news-article60480832 and AN EARTHQUAKE (1862, September 12). *The Argus* (Melbourne, Vic.: 1848–1957), p. 6. Retrieved 4 November 2019, from http://nla.gov.au/nla.news-article5721963.
9   Transactions of the Royal Society of Victoria, 1861 to 1864.
10 *A History of Victoria*, Geoffrey Blainey.
11 *The Chronicles of Early Melbourne*, Garryowen.
12 *Letter – CJ La Trobe to RC Gunn*, 17 April 1848.
13 *Edward Wilson, journalist and editor* Author: Hurst, John In: *Porter, Muriel (Editor). Argus: The Life and Death of a Great Melbourne Newspaper (1846–1957) Melbourne, Vic.:* RMIT Publishing, 2003: [24–29].
14 *Marvellous Melbourne*, Jill Roe, 1974.

Chapter 8

1   1862 'ROYAL SOCIETY OF VICTORIA', *The Age* (Melbourne, Vic.: 1854–1954), 18 November, p. 6, viewed 5 November 2019, http://nla.gov.au/nla.news-article154971248.
2   1862 'ROYAL SOCIETY', *The Argus* (Melbourne, Vic.: 1848–1957), 18 November, p. 6, viewed 5 November 2019, http://nla.gov.au/nla.news-article6481243.
3   Ibid.
4   1862 'ROYAL SOCIETY OF VICTORIA', *The Age* (Melbourne, Vic.: 1854–1954), 18 November, p. 6, viewed 5 November 2019, http://nla.gov.au/nla.news-article154971248.
5   1862 'THE METEORITE AT THE MELBOURNE UNIVERSITY', *The Argus* (Melbourne, Vic.: 1848–1957), 22 November, p. 5, viewed 18 November 2019, http://nla.gov.au/nla.news-article6481361.
6   1862 'THE BRUCE METEORITE', *The Argus* (Melbourne, Vic.: 1848–1957), 20 November, p. 6, viewed 18 November 2019, http://nla.gov.au/nla.news-article6481301.
7   1862 'THE BRUCE METEORITE', *The Argus* (Melbourne, Vic.: 1848–1957), 5 December, p. 7, viewed 16 November 2019, http://nla.gov.au/nla.news-article6481631.
8   1862 'THE VICTORIAN METEORITE', *The Argus* (Melbourne, Vic.: 1848–1957), 27 December, p. 5, viewed 20 November 2019, http://nla.gov.au/nla.news-article6482105.
9   Ibid.
10 Ibid.

11 Ibid.
12 1863 'THE BRUCE METEORITE', *The Argus* (Melbourne, Vic.: 1848–1957), 2 January, p. 6, viewed 2 December 2019, http://nla.gov.au/nla.news-article6482302.

Chapter 9

1. Formation of Impact Craters – Lunar and Planetary Institute (https://www.lpi.usra.edu/publications/books/CB-954/chapter3.pdf).
2. Letter, Sir Henry Barkly to McCoy, 1863/04/16, Mitchell library.
3. Letter, McCoy to Sir Henry Barkly, 1863/04/25, Museum Victoria, Outbound letter book 2, no. 63/81.
4. Letter, Sir Henry Barkly to McCoy, 1863/04/26, Mitchell library.
5. Ibid.
6. Board of Inquiry interim report – MV Archives National Museum of Victoria – Inwards Correspondence. Oldersystem 02626.
7. Ibid.
8. Letter, Sir Henry Barkly to Mueller, 1863/05/22.
9. Ibid.
10. Letter, Sir Henry Barkly to Maskelyne, 1863/05/25.
11. Ibid.
12. Letter, Maskelyne to Anthony Panizzi, 26 July 1863.

Chapter 10

1. Earth Impact Database.
2. Letter, Mueller to Evans, 1863/09/11.
3. Mueller to McCoy, 11 September 1863, Regardfully Yours.
4. Ibid.
5. Ibid.
6. Mueller to Maskelyne, 25 September 1863, NHM.
7. Mueller to Sir Charles Darling, 28 November 1863, PROV.
8. Mueller to Maskelyne, 25 October 1863, NHM.
9. Mueller to McCoy, 28 October 1863, Public Records Office Victoria.
10. McCoy to Mueller, 11 November, Melbourne Museum.
11. McCoy to the Chief Secretary, 12 November, Melbourne Museum.
12. Mueller to Sir Charles Darling, 28 November 1863, Regardfully Yours.
13. Ibid.
14. 1863 'VICTORIA', *Sydney Morning Herald* (NSW: 1842–1954), 25 December, p. 8, viewed 24 February 2020, http://nla.gov.au/nla.news-article13092561.

15 Museums Victoria Collections https://collections.museumvictoria.com.au/items/1449038. Accessed 04 February 2020.
16 Maskelyne to Mueller, 8 January 1864, Regardfully Yours.
17 Mueller to Maskelyne, 25 March 1864, NHM.
18 Mueller to Maskelyne, 25 April 1864, NHM.
19 Sir Charles Darling to Chief Secretary CO309/67, page 275.
20 Mueller to Maskelyne, 25 April 1864, NHM.
21 1864 'LEGISLATIVE ASSEMBLY', *The Herald* (Melbourne, Vic.: 1861–1954), 20 May, p. 3, viewed 4 April 2020, http://nla.gov.au/nla.news-article247589776.
22 1933 'WESTERNPORT (v.)', *The Australasian* (Melbourne, Vic.: 1864–1946), 7 January, p. 33, viewed 10 April 2021, http://nla.gov.au/nla.news-article141369080.
23 Chief Secretary (McCulloch) to Sir Charles Darling 20 June 1864, CO309/67, p. 278.
24 Sir Charles Darling to Edward Cardwell, 20 July 1864, CO309/67, p. 272.
25 Mueller to Maskelyne, 1 January 1865.
26 1865 'THURSDAY, JANUARY 19, 1865', *The Argus* (Melbourne, Vic.: 1848–1957), 19 January, p. 4, viewed 4 April 2020, http://nla.gov.au/nla.news-article5745234.
27 McCoy to the Chief of Detective Police, 21 January 1865.
28 McCoy to the Chief Secretary, 21 January 1865.
29 McCoy to J. Pierce, esq. (Govt. Keeper of Stores), 23 January 1865.
30 The Penllergare Observatory Birks, J.L. Antiquarian Astronomer, 2005, Issue 2, pp. 3–8.
31 Mueller to Maskelyne, 25 January 1865.

Chapter 11

1 Murchison and District Historical Society Inc. https://murchisonhistoricalsociety.wordpress.com/2007/09/12/murchison-meteorite/.
2 Ref: https://doi.org/10.1046/j.1468-4004.2003.45208.x.
3 Henry, D.A., 2003. *Meteorites and tektites*. In *Geology of Victoria*, W.D. Birch, ed., pp. 663–670. Geological Society of Australia Special Publication 23. Geological Society of Australia (Victoria Division).
4 *High molecular diversity of extraterrestrial organic matter in Murchison meteorite revealed 40 years after its fall.* Philippe Schmitt-Kopplin, Zelimir Gabelica, Régis D.Gougeon, Agnes Fekete, Basem Kanawati, Mourad Harir, Istvan Gebefuegi, Gerhard Eckel, Norbert Hertkorn Proceedings of the National Academy of Sciences February 2010, 107 (7) 2763–2768; DOI: 0.1073/pnas.0912157107.
5 1865 'TOWN NEWS', *The Australasian* (Melbourne, Vic.: 1864–1946), 28 January, p. 7, viewed 14 May 2022, http://nla.gov.au/nla.news-article138037616.

6. *Report of an Examination of the Meteorites of Cranbourne, in Australia; of Rowton, in Shropshire; and of Middlesbrough, in Yorkshire* Walter Flight Philosophical Transactions of the Royal Society 1882.
7. 1866 'THE BRITISH MUSEUM', *Tasmanian Morning Herald* (Hobart, Tas.: 1865–1866), 29 June, p. 3, viewed 8 August 2020, http://nla.gov.au/nla.news-article169045928.
8. Macleod, Roy, 'Passages in Imperial Science: From Empire to Commonwealth', *Journal of World History*, vol. 4, no. 1, 1993, pp. 117–150. JSTOR, www.jstor.org/stable/20078549. Accessed 10 May 2020.
9. Ibid.
10. *Sir Roderick Murchison on the Australian Gold-fields. To the editor of the Bendigo Advertiser,* 9 June 1860, Trove.
11. Mueller to Joseph Hooker, 24 October 1865, *Regardfully Yours*.
12. *Formation of Museums in Victoria,* Frederick McCoy, 1856.
13. *The Gold Fields and Mineral Districts of Victoria,* R. Brough Smyth, 1869.
14. *Report of an Examination of the Meteorites of Cranbourne, in Australia; of Rowton, in Shropshire; and of Middlesbrough, in Yorkshire,* Walter Flight Philosophical Transactions of the Royal Society, 1882.
15. https://archive.org/details/TheStrandMagazineAnIllustratedMonthly/TheStrandMagazine1896aVol.XiJan-jun.
16. 1876 'The Brough Smyth Enquiry', *Newcastle Morning Herald and Miners' Advocate* (NSW: 1876–1954), 5 May, p. 3, viewed 25 July 2020, http://nla.gov.au/nla.news-article136858841.
17. *The Gold Fields and Mineral Districts of Victoria,* R. Brough Smyth, 1869.

Epilogue

1. *The Cranbourne Meteorites*, A. B. Edwards & G. Baker, 1944. Memoirs of Museum Victoria.
2. 1927 'METEORITES.', *The Age* (Melbourne, Vic.: 1854–1954), 28 December, p. 10, viewed 20 June 2021, http://nla.gov.au/nla.news-article202286737.
3. *The Good Country. Cranbourne Shire,* Niel Gunson, 1968.
4. *Dynamics of the Cranbourne Meteorite Fall,* Peter F. Skilton 22nd National Australian Convention of Amateur Astronomers 2006.

Index

Abel, Augustus 20, 22, 25, 27, 31–2, 40, 42, 46, 49, 56–9, 64–5, 71, 74, 77, 81, 89, 91, 97, 103, 109, 115, 130–1, 133
Abel, Frederick 32, 40, 42, 46, 89
Abel, Johann Leopold 89
Aborigines Protection Board 135
Acclimatisation Society 128
Acraman crater 87
Adelaide Rift Complex 105
Aitken, John 53
American Museum of Natural History 64
Amundsen, Roald 134
Anderson, Samuel 6
Apollo 11 120
Asteroid Belt 11, 23, 36–7, 49, 137
asteroids 12, 23–4, 36, 49
Astrolabe 19
Aurora Australis 39
Australia Felix 20, 118
Australian Colonies Government Act, 1850 (Port Phillip) 85
Australian rules football 92

Bacon, George 144
Bakewell, John 8
Ballarat 3–4, 14, 20, 25, 27, 39, 45, 80, 133
Baratta meteorite 136
Barker brothers 55

Barkly, Sir Henry 9, 12–13, 15, 27, 50, 57–8, 61, 65–6, 72–4, 79–82, 87, 94, 97–103, 129–30, 133–4
Barlangi Rock 104
Barnes, Benjamin 67
Barringer crater 87
Barringer, Daniel 48
Barry, Sir Redmond 5, 26, 46, 77
Barwon River 53
Bass River 53, 55
Bass Strait 2, 17, 19
Bass, George 17
Batman, Henry 31
Batman, John 29–31
Baudin, Nicolas 18
Becker, Ludwig 37, 73
Benares 103
Bendigo 3–4, 16, 38, 66, 77
Bendigo Advertiser 25, 123
Bentham, George 61
Berthelot, Pierre 130
Black Thursday 2–3
Black, John 17
Bleasdale, Reverend John 120
Blundy, Bill 141
Blundy, Glenn 139, 141
Blundy, Merle 141
Bode, Johann 11
Bolaven Plateau, Laos 96
Boonwurrung 19, 55, 124–5
Botanic Gardens, Kew 33

Botanic Gardens, Melbourne 16, 35, 57, 98, 106, 111, 126
Bounty Scheme 14
Bourke, Richard 30
Bowen, William 18
Brisbane Ranges 39
British Museum 25, 32, 35, 40–2, 45, 51–2, 56–60, 64–6, 69–71, 73–4, 78–81, 88–90, 92–4, 97, 100, 102, 105, 107–11, 113–17, 121–3, 125, 130–1, 133, 136
British Museum trustees 41, 81, 89, 97, 100, 102, 115
Bruce, Alexander 24
Bruce, James 8, 22, 24–5, 28, 49, 51–3, 57–60, 65–7, 69–70, 72, 74, 78–9, 88–94, 99–102, 105, 107–10, 113–15, 123–5, 129–30, 132–3, 145
Bruce, Mary 24
Bruce, Thomas 124
Burke & Wills 15, 20, 28, 71–3, 94–5, 129

Calcutta 80, 122
Cambrian (period) 1, 43–4, 137
Cambridge University 39, 43–5, 122, 128
Cameron, Alexander 7–8
Campell, Thomas 5
Canyon Diablo 76
Cape of Good Hope 99, 133
Cape Otway 17
Cape Paterson 6–7, 49
Cape Schanck 17, 55
Cape York meteorite 64, 75
Cardinia Creek 54, 114
Cardwell, Viscount Edward 115
Carnarvon, Earl of 134
Carnmallum 55

Carrington event 39–40
Ceres 11–12
Chambers, Enoch 67–8
Cherry Creek 53
Chicxulub 48, 105
Chief Secretary of Victoria 98, 100–01, 109, 113, 116–17, 120, 129
Clarke, Reverend William 128
Clyde, Firth of 14
Clyde, primary school 138–41, 145
Clyde, rural hamlet 114, 132, 138–9
Cohen, Emil 143
Collins, David 18–19
Colonial Office 4, 9, 29, 82, 97, 103, 108, 115
Columba 24
Committee of inquiry 78
Condell, Henry 84
Cook, James 17
Cooper Creek 73, 94–5
Corinella 56
Corio Bay 19
Corrowong 56
Cosmopolitan Land and Banking Company 143
Cranbourne meteorite fragment No. 1 (Bruce meteorite) 13, 15, 21–2, 26, 28, 49, 52, 56–8, 60, 64–7, 70–4, 78–9, 81, 87–8, 93, 100, 109, 114–17, 120–5, 127, 129–30, 132–3, 135–6, 140, 142, 145–7
Cranbourne meteorite fragment No. 2 (Abel meteorite) 13, 15, 22, 25, 27, 32, 40, 42, 46–7, 64, 66, 73, 77, 80, 103, 109, 112, 115–16, 125, 127, 130, 132, 136, 140, 142
Crawford crater 105
Cretaceous (period) 105
Croker, A.R. 145

Crooke, William 89, 93, 114

Daintree, Richard 49, 51, 57–9, 67, 74, 77–8, 100, 129, 135
Dalrymple, George 135
Dandenong, creek 54
Dandenong, ranges 19
Dandenong, road 68
Dandenong, township 7, 67
Darling, Ralph 109
Darling, Charles 19, 61, 108–09, 112–16, 127
Darwin Crater 97
Devonian (period) 105
Dividing Range 19, 29, 41, 118
Donati, Giovanni 37, 117
Duke of Newcastle 81, 109, 115
Dutton, Francis 34

Eades, Richard 10
Earl Grey 85
Ebden, Charles 29–30
Ediacaran biota 1
Ellery, Robert 38, 80, 111
Emerald Hill 83
Endeavour 17
Eocene (period) 105
Esmond, James 3
Eumemmerring Creek 54
Evans, George 53, 77–9, 98, 102, 105, 107–08, 114

Faraday, Michael 42
Fawkner, John 29–31, 53, 87
Fisher, David 53
FitzGibbon, Edmund 7–9, 12–13, 15– 16, 22, 27, 51–2, 57, 74, 81, 87–94, 98, 105–06, 124, 129–30, 133, 142

Flagstaff Hill 16–17, 30, 37, 39
Flaxman crater 105
Flemington 28, 83
Flight, Walter 131
Flinders Ranges 1
Flinders, Matthew 17–18
Flora 36
Flora Australiensis 62, 123
Foord, George 27–8, 67, 70, 72, 74, 129–30, 135
Ford, Ramsay 97
Foy, William 40
Fragment phytographiae australiae 62
Francis Ridley 14
Franklin, Benjamin 93
Franz Joseph, Emperor of Austria 15, 82, 100
French Island 18
Friend, Matthew 122

Gardens House, Botanic Gardens 106
Gardiner, John 29, 53
Gawler Ranges 87
Geelong College 91
Geological Society of London 41
Geological Survey of Great Britain 126
Geological Survey of Victoria 45, 49, 51, 134
Gerzeh 75
Gibraltar 100
Gilbert, Johny 121
Gillbee, William 88
Gisborne 53
Glass, Hugh 55
Glen Huntly 14, 132
Gondwana 105, 137
Gosse's Bluff 97

Goulburn River 34, 118
Grant, James 17–18
Gray, John 66
Great Comet of 1858 (aka Donati's Comet) 37
Great Comet of 1861 37–8
Great Exhibition, 1851 76
Gregory, Augustus 35, 62, 72, 95, 126
Griffiths, Charles 53
Gulf of Aden 99
Gulf of Carpentaria 20–1

Haidinger, Wilhelm 100, 130, 134
Hall, Ben 121
Harbinger 17
Hargraves, Edward 3
Harrison, Anna 43
Haushofer, Karl 130
Hawdon, Joseph 29
Henbury 87
Henry, Dermot 141
Henty family 28
Hepburn, John 29
Heyne, Ernst 57–8, 60
Hiramaya, Kiyotsuga 36
Hoba meteorite 48
Hobart 16, 18, 122
Hobsons Bay 37, 83
Hoddle, Robert 30
Hooker, Joseph 35, 62, 122
Hooker, William 33–5, 62, 122, 125
Hotham, Sir Charles 45
Hovell, William 6, 19–20
Howitt, Alfred 94
Howitt, Richard 54
Howitt, William 2
Humboldt, Alexander von 16, 134

Hume, Hamilton 14, 19
Humffray, John 27
Hunter, Stanley 145
Hygiea 12

Impact, craters/sites/structures 24, 37, 49, 64, 75, 86–7, 95–7, 104–05, 119, 134, 138
Imperial Museum of Mineralogy, Vienna 74, 80, 82
Indian Ocean 99
International Exhibition of Agricultural and Industrial Products, 1862 26, 46, 74, 76, 81
Irvine, Hans 144

Jack Hills 1
Jackson, Samuel 53
Jackson, William 53
Jacksons Creek 53, 140
Jupiter 2, 11, 23, 49, 138
Jurassic (period) 105

Kananook Creek 53
Kangaroo Island 99
Kata Tjuta 97
Keilor 14, 140
Kelly, Ned 5, 44
Kilmore 19
King George's Sound 99
King, Philip Gidley 18
King, John 73, 94
Kirkwood gaps 23
Knight, John George 77
Koo Wee Rup swamp 54, 138
Kulin 21, 31

La Trobe, Charles 4, 10, 14, 34, 82, 84–5

Lachlan River 20
Lady Nelson 17
Lake Eyre 95
Lalor, Peter 45
Langhorne, Alfred 53
Langwarrin 25, 138, 142–5
Late Heavy Bombardment 49
Laverton 53
Legislative Council, New South Wales 84–5
Legislative Council, Victoria 4, 8
Lincoln, Abraham 37
Lineham, James 13–15, 22, 25, 124, 130, 132
Lineham, Jim 132
Lineham, Suzanne 22
Llewelyn, Thereza 117
Lonsdale, William 29–30, 53, 82
Lovering, John 119

Macadam, Dr John 81, 88, 91–2, 113–14, 122, 135
macadamia nuts 92
Madras 80, 122
Magnetic Observatory, Melbourne 38, 95, 100
Magnetic Observatory, Rossbank 16
Mahomed, Dost 73
Mallum Mallum 55
Malta 100
Manton family 55
Maribyrnong River 28, 53, 83
Mar-ne-bek 142
Mars 11, 24, 49, 75, 139
Marseilles 99
Maskelyne, Nevil Story 33, 41–2, 46, 56, 59, 65, 71, 73, 79–81, 89, 97–9, 102, 105, 107–08, 110–13, 117, 122–3, 126–7, 129–31
Maximillian, King 16
Mayes, Charles 90, 108
Mayune 7, 54–5
McCoy, Emily 44, 127
McCoy, Frederick 9, 15, 27, 42–7, 49–52, 57–60, 62, 64–8, 71–2, 74, 77–9, 81, 90, 92, 97–102, 106–13, 115–17, 120–3, 126–30, 135, 143, 147
McCoy, Frederick Henry 127
McCulloch, James 109, 113–17, 129
McKay, Hugh 8, 12–15, 20, 22, 25, 52, 55, 67–8, 88, 90, 123, 130, 132, 142, 144–5
McMillan, Angus 56
Mediterranean Sea 99
Meisner, Carl 34
Melbourne 2–6, 8–10, 14, 16–17, 24–5, 27–8, 30–1, 34–5, 38, 40, 42, 44, 46, 54, 57, 65, 68–9, 73, 77, 79–85, 88–9, 91–3, 99, 102–03, 109–10, 112, 117–18, 122, 126–7, 133, 138, 140, 143–4, 146
Melbourne Exhibition, 1854 5, 26, 35
Melbourne Incorporation Act 84
Melbourne Punch 77, 79, 90
Menindee Lakes 75
Mercury 37–9
Merrett, Samuel 5, 26
Meteor Crater 48, 75
meteorites 6, 13, 16, 21–2, 25–7, 31–3, 36–7, 39–40, 42, 46, 48–9, 51–2, 56–71, 73–82, 87–92, 94, 97–103, 105–10, 112–21, 123–5, 127, 129–32, 136–8, 140–7
meteoroids 24, 36–7, 49, 63–4, 137, 142
Mexico 48, 82
Milligan, Joseph 32–3, 35, 56, 125
Mitchell, Thomas 14, 20, 29, 96, 118
Mollison, Alexander 29

Moon, the 12, 37, 48–9, 75, 86, 110–11, 118–19
Moorabool River 53
Mordi Yallock waterway 53
Morgan, Dan 121
Mornington 7, 24
Morton, William 5
Mount Macedon 29
Mueller, Ferdinand 9–10, 15, 32–5, 40–3, 56–62, 64–7, 70–4, 78, 87–90, 92–5, 98–9, 101–02, 105–15, 117, 123–9
Mundrabilla meteorite 145–6
Mundy, F. M. 55
Murchison, Charlotte 41
Murchison, Sir Roderick 40–1, 43–4, 50, 56, 80–1, 123, 128, 137
Murchison meteorite 119–20, 137, 146
Murchison, town 118–19
Murray, John 18
Murray, Reginald 143
Murray, Robert 54–5
Murray River 19–20, 29, 35
Murrumbidgee River 20, 30

National Museum of Victoria 43, 46, 49, 51, 57–60, 66, 68, 74, 78, 100–01, 109, 115, 117, 128, 130–1, 143
Natural History Museum, London 131, 133, 146
Naturaliste 18
Nelson, James 8, 28
Neumayer, Georg 16–17, 20–2, 25, 38–9, 52, 58, 60, 67–8, 70, 72–4, 81, 100, 111, 124, 129–30, 134
New Zealand 28, 79, 98, 135
Norfolk 17
North Australian Exploring Expedition, 1855 35, 95

Nullarbor Plain 75, 96, 145

O'Brien, Henry 83
O'Connor, Terence 54
O'Shanassy, John 77, 109
Owen, Richard 33, 59, 65, 71, 73, 77, 79–81, 89, 107, 112, 117, 121, 123, 127

Padley, Alfred H. 143–4
Palaeozoic (period) 43–5, 49
Pallas 12
Park Forest meteorites 138
Parnallee meteorite 123
Patterson, Alexander 114, 138
Pearcedale 142, 144–5
Peninsular and Oriental Steam Navigation Company 99
Permian (period) 105
Philosophical Institute of Victoria 9–10, 35, 45, 61, 72
Philosophical Society of Victoria 9, 35
Piazzi, Giuseppe 11
Point de Galle, Ceylon 99
Port Jackson 17–18
Port Phillip Association 28, 53
Port Phillip District 2, 4–5, 14, 17–21, 28–31, 40, 53–4, 82–5, 103, 110
Portland 2, 17, 28
Prince Albert 73
Prodromus of the Palaeontology of Victoria 128
Prodromus of the Zoology of Victoria 128
Pueblito de Allende 119

Queen Victoria 5, 32, 84

Red Rover 117, 120

Red Sea 99
Red Tuesday bushfire 144
Reed and Barnes 10
Reichenbach, Reverend Carl von 131
Richardson, Maurie 138, 140–1
Richmond Paddock 111
Ridgway, Charlotte 14
Rossbank Observatory 16
Rowe, George 17, 77
Royal Park 15
Royal Society of London 40–1, 57, 112, 131
Royal Society of Victoria 9–10, 12, 15–16, 22, 27–8, 32, 58, 60–2, 72–4, 77, 81–2, 87, 89, 91–2, 94, 102, 105, 107–08, 112, 114, 120, 122, 125, 129, 135
Ruffy brothers 54–5
Ruffy, Frederick 54
Rupprecht, Karl 25
Russell, George 53

Sabloniere Hotel 25
Sadler, Gerard 139–41
Sargeaunt, William 115
Schliemann, Heinrich 124
Scotch College 91
Scott, James 6, 27
Secretary of State for the Colonies 56, 81, 115, 134
Sedgwick, Adam 41, 43–5, 77, 128, 137
Selwyn, Alfred 6–7, 9, 27, 47, 49–51, 59–60, 65, 67–8, 70–2, 74, 77–8, 90, 93, 100, 102, 122–4, 126, 128–9, 134
Shark Bay 1, 87
Shergotty meteorite 123
Sherwood 8, 12, 24–5, 132, 144–5
Sherwood Park 25, 133, 144

shocked quartz 105
Silurian System 41, 43–4
Skeleton Creek 53
Sloane, Hans 32
Smith, Alexander 88
Smith, Jennet 14
Smith, Robina 14
Smith, William 55
Smithsonian Institute 144
Smyth, Robert Brough 77–9, 88, 101–02, 129–30, 134–6
Snowy Mountains 55
Solar System 1, 11–12, 24, 48–9, 64, 119, 138, 142
Solomon, Joseph 53
Sonder, Wilhelm 62
South Australia 1, 18, 33, 40, 72, 87, 105, 126
Squatting Act, 1836 53
St Germains 114, 138
Stawell, Sir William 9–10
Stephenson, George 93
Stieglitz, Robert William von 53
Strachan, James 53
Strutt, William 2
Strzelecki, Pawel 14, 41, 55–6
Suez 99
Sullivan Bay 18–19
Sunbury 53, 77
Supreme Court 83
Swainson, William 34
Swift-Tuttle comet 80
Switzerland 82
Sydney Morning Herald 38

Talundilly 87
Tarra, Charlie 56
Tasmania 2, 18, 28, 33, 96, 141

Tebbutt, John 38, 80
tektites 96
Age, The 5, 89, 110, 143
Argus, The 5, 24, 38–9, 51–2, 81, 85, 89–90, 92, 95, 102, 111, 113, 115, 120
Herald, The 31, 68
Strand Magazine, The 131
Theia 49
Thorsberg meteorites 137–8
Thorsberg quarry 137
Titius, Johann 11
Tnorala 97
Todd, Charles 38–9
Tookoonooka crater 87
Toomuc Creek 54
Tooradin 8, 15, 55, 144
Towbeet 7, 55

Uluru 97
United States/America 20, 23, 42, 47, 51, 64, 82
University of Melbourne 5, 9, 44, 46, 68, 91, 115, 119, 127, 131
d'Urville, Jules 19

Van Diemen's Land 17–18, 20, 28–9, 31–2, 53–4, 83, 122
Venus 39
Vesta 12, 36
Victorian Exhibition, 1861 26, 31, 46, 67, 77
Victorian Exploring Expedition 15
Victorian Institute for the Advancement of Science 9, 35
Viennese Royal Academy of Sciences 130

Walcott, Richard 131, 143

Wallace, Alfred 37
Warburton Basin, the 95
Warburton craters 105
Warburton, Peter 72
Warrigal Creek, massacre 56
Webster, Sam 55
Wedge, John 53
Wellington 79
Werribee River 53
Western Port 2, 6–7, 17–19, 27, 42, 53–5, 144, 147
Wheeler, Dianne 24
White, Edward 37–8
Widmanstatten pattern 28, 32, 76, 147
Wil-im-ee Moor-ring (Mount William quarry) 21
Williamstown 84
Williamstown Observatory 80
Wilson, Edward 85
Wilson, William 9
Wilsons Promontory 18, 34
Woiwurrung 30
Wolfe Creek 87
Woodleigh dome 87, 104–05
Woodwardian Museum 43
Wurundjeri 29

Yallock Creek 53
Yarra River 4, 8, 21, 29–30, 53, 83–4, 110–11, 133
Yarrabubba 104
Yass 55, 83
You Yangs 37

Zimmerman, Karl 130
zircon 1, 104
Zoologic Gardens, Melbourne 35, 111

www.ingramcontent.com/pod-product-compliance
Lightning Source LLC
Chambersburg PA
CBHW020859020526
44116CB00029B/865